Changing the Classroom from Within

Partnership, Collegiality, Constructivism

A guide and reference book for professional
development and improvement of the
teaching-and-learning process

Joseph I. Stepans
Barbara Woodworth Saigo
Christine Ebert

SP
SAIWOOD PUBLICATIONS
Montgomery, Alabama
ISBN 0-9649967-0-7
1995

For further information about the WyTRIAD professional development model, including getting a program started, contact:

Dr. Joseph Stepans
Professor of Science and Mathematics Education
University of Wyoming
P. O. Box 3992
Laramie, WY 82071

Published and distributed by

SAIWOOD PUBLICATIONS
P. O. Box 242141
Montgomery, AL 36124

Telephone: (334) 277-3433
Toll-free: (800) 743-4787
Fax: (334) 277-0105
  email: bsaigo@aol.com

ISBN 0-9649967-0-7

Suggestions and corrections are welcomed.

The activities, strategies, and models presented in this book are consistent with **national standards** *for learning, teaching, and professional development that have been established by the National Council of Teachers of Mathematics, National Research Council, and National Staff Development Council.*

Printing and binding by
WALKER PRINTING
Montgomery, Alabama

Printed on recycled paper

Dedication

To our children, their contemporaries, and future generations of children who will seek to understand and participate constructively in our world.

With thanks to the teachers who continue to grow
in their understanding of learning.

The Authors

Joseph I. Stepans is Professor of Mathematics and Science Education at the University of Wyoming, in Laramie. He has a B. S. Degree in physics and mathematics, M. S. Degree in physics, and Ph.D. in science education. Dr. Stepans has done research on children's views of the world and on teaching and learning strategies since 1981. He has worked actively to bridge his own research and the research of others into the classroom. His focus on finding more appropriate ways to help students learn is an outgrowth not only of his interest in research, but also of his childhood experiences and his years of teaching physical science, physics, chemistry, and mathematics at the precollege level. This book represents the current synthesis. Dr. Stepans has shared his ideas and models with thousands of teachers and other educational professionals through workshops, conferences, and publications.

Barbara Woodworth Saigo is a biologist, author, and independent educational consultant. She has a B. A. Degree in biology and M. A. Degree in zoology (emphasis in natural history and ecology). While pursuing her master's degree, she completed requirements for secondary biology teaching and is presently in a science education Ph.D. program. In addition to co-authoring textbooks in botany, biology, and environmental science, she has taught university courses in biology, botany, zoology, and ecology, including biology courses designed for pre-service teachers and adult outreach courses. She has directed Eisenhower and NSF-supported in-service projects. Most recently she was Director of Sponsored Research and Contracts at Southeastern Louisiana University.

Christine Ebert is Associate Professor in Instruction and Teacher Education at the University of South Carolina, in Columbia. She has a B. S. Ed. degree in biology, M. S. in secondary education, and Ph.D. in science education. While her primary interest is in helping elementary teachers learn how to enjoy the teaching of science, she has spent much of her time working side-by-side with teachers in professional development schools. The hundreds of interviews she has conducted with children and teachers collectively produce a first-hand understanding of the ideas children bring with them into the classroom.

Acknowledgments

We offer our sincere appreciation to the following colleagues who assisted us by reviewing the Field Test Edition.

Diane Galloway, Director
Wyoming Center for
Teaching and Learning
University of Wyoming
Laramie, Wyoming

Carol Marino
Principal
Osceola Magnet School
Vero Beach, Florida

JoLane Hall
Classroom Teacher
Pontiac Elementary School
Elgin, South Carolina

Diane Schmidt
Classroom Teacher
Franklin Park Magnet School
Fort Myers, Florida

Carole J. (Dody) Kinney
Classroom Teacher
Lincoln Elementary School
Torrington, Wyoming

Toni Sindler
Principal
Franklin Park Magnet School
Ft. Myers, Florida

Ted Kinney
Classroom Teacher
Trail Elementary School
Torrington, Wyoming

Stella Wilkins
Classroom Teacher
Pontiac Elementary School
Elgin, South Carolina

Melissa Klosterman
Classroom Teacher
Pontiac Elementary School
Elgin, South Carolina

Sharon Williams
Classroom Teacher
Pontiac Elementary School
Elgin, South Carolina

Dr. Ivo Lindauer
Department of Biological Sciences
University of Northern Colorado
Greely, Colorado

Dr. Donna Wolfinger
School of Education
Auburn University at Montgomery
Montgomery, Alabama

*Learn about a pine tree from a pine tree,
and about a bamboo plant from a bamboo plant.*

Matsuo Basho
Japanese poet
1644-1694

Δ

Table Of Contents

Topics in depth
Collecting data about the effectiveness of teaching and learning
through classroom observation
What kind of data do I need to collect?
How do I collect data?
Where do I record my observations?
How can this recorded information be used?
Peer coaching
What is peer coaching and when do you use it?
How does it work?
What is necessary to insure successful peer coaching?
Between-the-session activities
Interaction of the partners
Specific notes for teachers
Specific notes for administrators
Specific notes for the facilitator

Chapter 9   Session IV Activities and Topics   147

About the session
Activities
Share results of classroom observations and peer coaching
Integrate colleagues ideas to explore in own classroom
Integrated CCM lesson is modeled
Discussion and reflection
Topics in depth
Creating curriculum that is appropriate and accessible for students
Examining curriculum
Questions to ask when examining the degree to which curriculum,
instruction, and assessment are aligned
Between-the-session activities
Interaction of the partners
Specific notes for teachers
Specific notes for administrators
Specific notes for the facilitator

Chapter 10   Session V Activities and Topics   157

About the session
Activities
Share results of classroom observations and peer coaching
Share results of teaching integrated lesson
Share results of reviewing district curriculum
Discuss impact of national criteria on curriculum
Discuss next step
Discussion and reflection
Topics in depth
How can our expectations of students, the instructional experiences
we provide, and assessment be aligned?
Expectations
Experiences
Assessment
How to share and extend your experiences
Looking ahead

Prologue

It is the teacher who, in the end, will change the world of the school by understanding it.

...Lawrence Stenhouse

What do we want to accomplish?

If we accept the philosophy expressed by Lawrence Stenhouse, then we need to insure that the classroom teacher possesses the **expertise** and the **knowledge** and the **support** and the **time** to implement effective teaching practices in his/her own classes. The models presented in this book seek to do several things.

- Concentrate on creating meaningful, systemic change.

- Provide teachers with skill, time, and support to take ownership of curriculum, instruction, and assessment.

- Create an environment where teachers conduct practical and meaningful research.

- Have teachers use their own research to judge the appropriateness of curriculum, instruction, and assessment.

- Increase the number of outstanding teachers.

- Have teachers develop a constructivist view of both teaching and learning.

- Have teachers align curriculum, instruction, and assessment to meet the needs of all their students.

1

What is WyTRIAD?

WyTRIAD is a **structural model for professional development**. It synthesizes research about the teaching-and-learning process into practical application (see Figure). Embedded within the WyTRIAD is a constructivist instructional model that fosters conceptual change.

In the WyTRIAD, the classroom teacher, building administrator, and professional development facilitator develop a long-term partnership to create an environment that is dedicated to (1) examining what is happening in the classroom and (2) improving the learning experiences of students. Each partner makes a commitment and actively contributes to the process.

The WyTRIAD is an effective vehicle to provide teachers with expertise, knowledge, support, and time in their quest to make their classrooms more productive and effective learning places for their students. It is a powerful, tested, successful model that has been implemented in Wyoming, South Carolina, and Florida.

**Features that are integrated into
the WyTRIAD in-service model**

Δ

sharing
modeling
interviewing
peer   coaching
collaborative   learning
keeping reflective journals
development   of   leadership
teaching   for   conceptual change
preparing and teaching sample lessons
changing views about teaching and learning
keeping track of changes in students and selves
making   classroom   observations   and   collecting   data
teacher as a researcher, doing classroom-based   research
identifying meaningful   changes in teachers and in students
reviewing, evaluating, and making decisions about curriculum
aligning expectations of students, instructional experiences, and assessment
full partnership of teacher, administrator, and professional development person

---

Who are the partners?

The teacher

Instead of following a prescribed curriculum and allowing the curriculum to evaluate her/his performance and that of the student, the teacher becomes the judge of the curriculum and instruction. The teacher uses the results of his/her **own observations** of students and colleagues to reflect on what to teach, how to teach, what to assess, and how to assess meaningful changes that may be taking place in the student. By participating in the WyTRIAD sessions, the teacher receives training in **classroom-based research** and the application of a **conceptual change model** of teaching and learning.

The building administrator

As an active partner involved in discussions and activities, the administrator comes to understand the professional needs of her/his teachers and is willing to provide the necessary support and time for them. By participating in the WyTRIAD sessions, the administrator also learns about classroom-based research and the application of a conceptual change model of teaching and learning. The administrator provides leadership to teachers in order for teachers to take risks in using non-traditional ideas in their classes. The support of the administrator is <u>essential</u> for creating needed change.

The professional development partner

The professional development partner (a university faculty member or an educator with WyTRIAD experience) provides expertise with (1) learning from children and (2) using children's ideas as a basis for instruction. This person assists teachers and administrators with implementing effective teaching practices, ideas for collaboration among teachers, conducting and interpreting classroom observations, and using classroom-based data to make appropriate changes in what to teach, how to teach, and what and how to assess. In this book, we refer to this partner as the professional development **facilitator.**

The team

In schools where a true partnership has existed and all the partners have honored their commitments, teachers have been empowered to use the teaching strategies, curriculum, assessments and materials that best serve the children in their own classrooms.

Who will find this book useful?

The purpose of this book is to share some experiences, information, and ideas that will enable teachers to develop and practice skills to improve both teaching and learning. The book is based on a model that was initiated in Wyoming; hence the name, WyTRIAD. The book is designed to be as useful as possible to several kinds of users.

- Teachers and staff development persons may find it useful to help them teach and make curriculum decisions more effectively, as well as to help them become leaders in doing in-service activities with their colleagues.

- The book also can be useful to university faculty, state department of education representatives, school administrators, researchers, and curriculum designers.

- University students in both pre-service and graduate programs can benefit from understanding these ideas and activities, which provide concrete examples of teaching that: (1) takes into account the present national calls for reform and the highly influential national standards in the disciplines, and (2) translates research into practical application.

- Partners involved in a WyTRIAD experience will find it to be a helpful text and reference book.

Most of the examples we have used focus on science and mathematics instruction. The principles and the process, however, are applicable to other subject areas, such as social studies, and there is a strong language arts potential.

PART ONE:
The Teaching-and-Learning Process

Chapter 1
Why is the educational system
changing so slowly?

Chapter 2
How can problems be addressed?

Chapter 3
How does WyTRIAD incorporate
a constructivist approach?

Chapter 4
What is meaningful restructuring of
the teaching-and-learning process?

6

Chapter 1

Why is the educational system changing so slowly?

The nature and extent of the individual's prior knowledge as well as contextual factors determine the nature of the knowledge which is constructed.   Because all new knowledge is filtered through the framework of beliefs which the teacher already possesses and is adapted to fit those existing frameworks, simply giving the teacher new curriculum or suggesting changes in practice may not result in the desired outcomes.

... C. Briscoe, 1991

Introduction

Higher education faculty members, teachers, and school administrators have worked in an atmosphere that proclaims "educational reform" for more than 20 years. During this time, catch-phrases associated with various efforts have occasionally caught media and, therefore, public attention, for better and for worse. "New Math" was a highly publicized experimental program that, unfortunately, created a long-lasting negative public perception of educational reform. Since then, ideas such as "learning contract," "hands-on learning," "computer-assisted instruction," "process writing," "reading and writing across the curriculum," "collaborative learning," "cooperative learning," "outcome-based education," and others have rippled through the educational community.

These ideas have been joined, most recently, by the national movement toward "scientific and mathematical literacy," "standards" in all subject areas, and national educational goals known as *Goals 2000*. **It seems particularly clear** that the standards and related recommendations of the national professional organizations and associations will, in fact, encourage and guide changes in science and mathematics education, immediately and for the foreseeable future.

Not all parents, school boards, community members, legislators, and professional educators are accepting of the present reform ideas. Among many, there is suspicion that the present move to reform is just another fancy idea being pressed upon society, when what children really need from the schools is uniform discipline, reinforcement of moral and religious standards, and a return to a traditional, "back-to-basics" approach. In some areas, conservative grass-roots organizations have been created to actively oppose educational reform, which they perceive as part of a subversive liberal agenda.

Usually, children are the losers in this debate, especially those children (the majority) who do not learn best by the "old-fashioned" lecture, note-taking, textbook-based, rote style of instruction. It is not our purpose here to engage in debate, however. **Our purpose here is to share specific, useful strategies to address the problems faced by children and teachers in the classroom.**

In this chapter we examine why our classrooms are not changing, or are changing so slowly. Is it because educational research is not reaching the classroom? because in-service experiences are not effective? because we are not paying attention to what is really happening in the students in regard to the concepts we are trying to help them develop? We think all of the above are major factors in the problem.

Is it because research is not reaching the classroom?

What is the role of educational research?

Just as biology, chemistry, physics, astronomy, and psychology need research to advance understanding about these fields, so does education. In education, research about how learning occurs, the characteristics of the learner, and the effectiveness of specific educational materials and practices can be useful for improving the learning environment, the curriculum, the experiences provided for the student, and the assessments.

Research on how learning occurs and how best a teacher can encourage active learning forms the basis for both the WyTRIAD in-service model and the teaching for conceptual change instructional model. Research on the following questions is considered during the WyTRIAD and in this book.

- Why hasn't the research reached the learner?

- What are the inadequacies of the traditional one- or two-day in-service model?

- What have we learned about the kinds of in-service experiences that are truly effective in helping teachers implement what they have experienced?

- Why is it important to confront and deal with the misconceptions that both students and teachers bring to the classroom about specific science and mathematics concepts and how do we accomplish this in a dynamic, non-threatening way?

- Why is it important to use students' experiences and preconceptions as a basis for teaching specific science and mathematics concepts?

- How can teachers unleash the power of the conceptual change model of learning in their day-to-day teaching?

- Why don't most existing assessment methods adequately assess learning, and how can assessment be designed to measure conceptual change and understanding?

What does educational research tell us?

Recent educational research has provided many potential keys to our puzzles about the ways students learn and perform, particularly in science and mathematics. Some significant points are summarized below. Many of them are elaborated more fully in subsequent chapters. For more detailed information about many of these points, you may want to review the list of references.

About learning?

The following statements summarize some of the things we are learning about learning.

- Piaget's description of intellectual development is valid (Figure 1.1).

- It is important to understand the sequence of stages in the development of operational thinking when planning learning activities.

9

- Thinking experience is more important than memorization of symbols.

- Learners must frequently manipulate materials to develop concepts.

- The understanding of concepts must come first, while words and symbols are added later to facilitate communication.

- Students are not blank slates, but come to the classroom with pre-formed concepts and explanations about the world.

- For many students, "real world" knowledge and "classroom" knowledge are different.

- Children's views of the world may be different from adult views.

- Many learners keep their views even after instruction.

- It is not that children do not understand, but that they understand differently.

- Children's misconceptions may be similar to those held by people a long time ago.

- If a concept does not address learners' naive ideas, the learners will resist it and hold on to their own ideas.

- There are many contributing factors to students' retention of misconceptions.

- It is necessary to use varied, alternative approaches to help learners to overcome their misconceptions.

Cognitive Levels
Data expressed as % of students sampled

AGE	PREOPER-ATIONAL	CONCRETE ONSET	CONCRETE MATURE	FORMAL ONSET	FORMAL MATURE
5	85	15			
6	60	35	5		
7	35	55	10		
8	25	55	20		
9	15	55	30		
10	12	52	35	1	
11	6	49	40	5	
12	5	32	51	12	
13	2	34	44	14	6
14	1	32	43	15	9
15	1	14	53	19	13
16	1	15	54	17	13
17	3	19	47	19	12
18	1	15	50	15	19

Figure 1.1. Studies such as this one by Epstein (1979) and those by Renner and Marek (1988) and others continue to demonstrate the validity of Piaget's view of cognitive development.

About what and how we teach and assess?

Summarized, the research on teaching and learning reveals the following profiles (Figures 1.2, 1.3, 1.4, 1.5)) of what are our **current, predominant practices** (left column), especially in teaching mathematics and science. These practices are contrasted to what research shows **should be happening** (right column).

What do we teach?	What should we teach?
•Isolated facts and symbols	•Concepts and facts in context
•Topics needed to prepare for the curriculum of the next class or grade	•Skills •Applications •Creative and imaginative perspectives •Positive attitudes toward the subjects and learning in general
•Information that lacks relevance	•Topics dealing with major concerns and concepts rather than trivia

•Information that lacks connections	•Relevance and connections to other concepts and knowledge and to learners' "real world"
	•Scientific literacy, leading to a habit of addressing such questions as
	   » What is the evidence?
	   » How do we know?
	   » Why do we believe?
	   » How do we find out?

•Elementary science that is modeled after middle school science modeled after secondary science modeled after college science	•Concepts that are appropriate to the readiness and needs of the learners

Figure 1.2 Predominant curriculum emphases compared to recommendations based on educational research and national standards.

How do we teach?	How should we teach?
•Depend mainly on lecture and reading	•Accommodate various learning styles
•Use instructional materials that lack relevance	•Use children's ideas as a basis for instruction
•Do laboratory experiences after the fact, for confirmation	•Develop a constructivist view of learning
•Emphasize "coverage"	•Emphasize <u>what is learned</u> versus <u>how much is covered</u>
•Emphasize fragmented ideas and concepts	•Emphasize understanding, making connections, and application of knowledge
•Reward the "right" answer	•Encourage a questioning environment
	•Create an environment of contradictions and inconsistencies to challenge existing preconceptions and motivate learning

Figure 1.3. **Predominant teaching practices compared to recommendations based on educational research and national standards.**

What do we evaluate?

- Verbal skills, especially reading and note-taking

- Ability to memorize from external sources and low-level recall of facts, definitions, symbols, and formulas

- Overemphasis on results

- Student performance only

What should we evaluate?

- Development of concepts

- Integration and application of knowledge
- Creative and imaginative approaches
- Skills
- Thinking
- Intellectual development
- Conceptual change
- Attitudes
- Problem-solving skills

- Total learning process

- Instruction, including the appropriateness and effectiveness of
  » curriculum content
  » lectures and demonstrations
  » laboratory experiences
  » books
  » teaching strategies
  » questioning practices
  » attention to children's ideas

Why do we evaluate?

- To assign grades
- For grade and course placement
- For advancement
- For school and district statistical purposes

Why should we evaluate?

- To assess learning
- To identify individual strengths, weaknesses, and differences
- To modify curriculum, instruction, and assessment

Figure 1.4. Predominant assessment practices and rationale compared to recommendations based on educational research and national standards.

How do we evaluate?	How should we evaluate?
•Written quizzes and exams •Standardized tests	•Provide students a variety of ways to exhibit learning •Provide opportunities for students to apply new understanding of concepts and skills •Observe performance •Interview to assess conceptual change •Listen to questions and responses •Create rubrics
•Machine scoring	•Understand limitations of machine-scored exams in detecting meaningful change

Figure 1.5. Predominant evaluation methods compared to recommendations based on educational research and national standards.

What are the limitations of traditional educational research and why is it inaccessible to teachers?

A major characteristic of scientific research is that investigators are required to design experiments and studies in which most factors (variables) are controlled (kept constant) so the effects of one or a small number of variables can be tested. It is also required that research results can be replicated and verified by others in similar situations. In the natural sciences it is usually possible to identify and control the variables _and_ subsequently be able to generalize the findings.

In educational settings, variables are more difficult to identify and control. Because of the demand to be as scientific as possible, most of the research on teaching and learning has been conducted in contrived situations or on topics of very narrow focus by educational researchers and graduate students at higher education institutions. The traditional, minimalist design of educational research is important for identifying statistical patterns and relationships--as in the analysis of test items, demographic information, and questionnaires.

This kind of design may result, however, in conclusions that are difficult to apply in a practical way to classroom situations. Whereas experimental research and statistical analysis demand elimination of variables, a given classroom on any given day is a stew pot of simmering variables in unpredictable states of flux. Because they have eliminated this total milieu--the context of the classroom--many research findings are of limited relevance and limited usefulness to the classroom teacher. A further reflection of the impact of context on research is that published research results sometimes even conflict with one another.

Limited accessibility of research findings to teachers is another problem. Research results are typically communicated among professional researchers, in professional publications and at professional meetings, using research language.

Furthermore, when research results are collected and communicated to teachers, it is often in the form of ideas and recommendations that are left to the teachers to deal with and figure out how to implement and integrate. The mechanism for learning how to effectively use the recommendations is missing.

Finally, research has had a limited impact on classrooms because teacher involvement has been very limited. While educational studies have used or manipulated teachers and instructional techniques, the teachers themselves have not had a collaborative role in the design of the studies.

Because of all of these factors, very little of what educational researchers find is implemented by teachers. As a result, many classroom practices remain unchanged and unaffected by research findings.

What is classroom-based research?

By contrast, **meaningful changes are likely to occur in classrooms where the classroom teacher conducts research in her/his own classroom.** There is now a popular trend toward more naturalistic, contextual research, especially research that involves classroom teachers as investigators (researchers) of their own students and teaching practices rather than just as objects of study by other educational professionals. This type of research is most commonly done as a collaboration between university researchers and classroom teachers.

The perspective of a teacher making classroom observations is different from one from outside the school setting. One of the differences is that teachers get immediate, spontaneous, often unpredictable feedback, whereas an external researcher has to intrude into the classroom in some way to collect feedback. Sometimes the design for collecting the feedback is deliberately limited and rigid, which may be good for testing a research hypothesis, but insensitive to the moment or the context. An external researcher tries to design research to control or ignore certain things, but the teacher deals in the world of spontaneity.

16

Perhaps a good analogy is that the teacher lives inside the fish bowl instead of looking at it from outside.

What is "not research?" Teachers often make adjustments based on student responses to materials, activities, assignments, or quizzes. Typical classroom observations and unexamined assumptions may seem useful, but they also can be misleading and simply not provide enough information upon which to base decisions.

Classroom research is more methodical than a casual gathering of observations and impressions (Chapters 5 and 6). It also produces a written body of data upon which to base conclusions. Teachers seldom conduct **organized, systematic inquiries** (research) on what is happening in the classroom. Perhaps this lack of research by teachers reflects a gap in our teacher education programs and also underscores the need for and importance of continuing professional development. What are some examples of useful classroom research?

> • Classroom research is being conducted by the teacher when the teacher interviews students individually or in small groups to find out first-hand what they think, know, and feel.
>
> • The teacher is conducting research when he/she is purposefully observing students, recording and analyzing (reflecting upon) the observations.
>
> • A teacher is conducting research when she/he is implementing a new teaching strategy and assessing its effectiveness on students' understanding and attitudes.
>
> • A teacher is conducting research when he/she is critically examining the curriculum in light of what she/he has learned from talking with students, observing students, and making appropriate revisions in curriculum, instruction, and assessment.

Is it because in-service experiences are not effective?

Ongoing professional development is both a requirement and an opportunity for continued learning, rejuvenation, and growth. Unfortunately, staff development workshops are often not regarded by the teachers as practical or relevant, nor do they have an enduring impact. Ineffective in-service experiences are not likely to change teacher behavior, let alone the school culture.

Characteristics of **ineffective in-service experiences** are well documented. The following statements are derived from numerous sources.

- Lacking a solid conceptual model, a staff development program is typically a series of unrelated topics.

- Teachers do not select the topics, topics are chosen for them.

- "One-shot" staff development workshops do not change teacher behavior.

- Teachers receive in-service experiences like a "dose" of medicine without being provided an opportunity over time to practice, reflect on how to incorporate them into student experiences in a meaningful way, analyze how it went and what to adjust, and construct their own knowledge.

- There is no follow-up to support teachers as they attempt to implement the experiences, to get questions answered, to obtain additional information or advice, or to share their successes and frustrations.

- Administrators, school boards, or departments of education "diagnose and prescribe" what they think teachers need, then evaluate and supervise the teaching even though they are not as close to the process as the teachers.

- There is a focus on technical models for the teaching process rather than on the dynamics of the learning process. As a result, teachers are forced to lock in on fixed procedures, models, and checklists instead of using and developing their own skills and creativity.

- The presentation is didactic rather than interactive, and the presenter is the "expert" who provides and controls the standardized topic, content, procedure, objectives, and outcomes.

- There is a narrow, non-individualized construct, in which the assumption is that all participants will be served equally, when the format that is chosen is usually the one that is most comfortable for the presenter.

- There is an emphasis on "training" rather than "educating."

- Particularly at the elementary level, workshops demonstrate many "fun" activities and attention-getters, but do not relate these sufficiently to fundamental concepts or purposeful investigations.

- Particularly at the secondary level, workshops on specific subject areas focus on content-related information (e.g., genetics, laws of motion, DNA electrophoresis, algebra) but do not model effective teaching and assessment strategies along with the content.

- Teachers are treated as clients or consumers, not as professionals with skills and knowledge--expertise that qualifies them to provide leadership and make decisions in the classroom and in the overall process of educational reform.

- Teachers lack administrative support to provide time and flexibility to implement new ideas and to work together.

Numerous authors also have discussed characteristics of what **effective in-service experiences** should include.

- Teachers should be involved in needs assessment, planning, designing, implementing, and evaluating the program.

- Long-term is better than short-term or "one-shot" programs.

- Experiences should be diverse, flexible, and meet the concerns of the teachers.

- Experiences should be readily usable.

- There should be a theoretical basis and rationale for teaching models that are presented, rather than just a "bag of tricks" approach.

- Subject area content should be linked to instructional strategies.

- The program should include modeling (demonstrations) by persons who are expert at using the teaching model.

- It should provide for practice and feedback in a low-risk (protected) environment, with opportunities for peer coaching.

- There should be a well-organized and responsive governing mechanism.

- There must be an evaluation plan that analyzes the individual in-service experiences <u>and</u> the overall program. Criteria should include both the quality, appropriateness, and effectiveness of the experience <u>and</u> evaluation of its subsequent impact on the classroom.

- Teachers should be provided time to apply what they've learned, gather data, get feedback from other teachers, reflect, integrate what they experience, and construct their own knowledge about teacher change. This necessitates a coherent, serial approach to staff development, with multiple related sessions <u>and</u> it requires administrative support.

- Meaning takes time and effort to construct. Opportunities for both self-reflection and sharing among peers should be a deliberate part of in-service design.

- Workshop leaders should be available for follow-up advice and assistance.

These and other characteristics of **effective** in-service experiences are incorporated into the WyTRIAD model. Characteristics that tend to interfere with teacher implementation of research findings and in-service experiences are avoided or minimized.

- Unlike most in-service experiences, WyTRIAD takes place over 4-5 months, during the academic year. It consists of several <u>sessions</u> in which the professional development facilitator and other resource persons (such as graduate students) meet and work with the teachers and administrators.

- The sessions are separated by <u>intervals</u> of several weeks during which teachers implement the ideas, strategies, and activities in their own classrooms.

- During the sessions, all partners have experiences and ideas to contribute--discussing what they have done, what they have observed, what they plan to do next based on these observations, asking questions, and dealing with directly relevant and applicable information, issues, and strategies.

- It has a high degree of teacher input and control, rather than being a "top-down" program.

- Research on the teaching-and-learning process is made accessible through its translation into practice and through involving teachers in their own research.

- Content information is linked to effective implementation strategies and the emphasis is on the process.

- All aspects of the experience are modeled for the participants, rather than just being discussed or presented.

- It occurs on-site, in the teachers' own communities and in their own classrooms, with their own students.

- There is readiness to change because the teachers choose the concepts around which they build their experiences.

- Teachers find that the changes they are being asked to implement are practical and make sense in the context of their classrooms, creating a high level of self-motivation for implementation.

- Based on their own research with their students, teachers make decisions about their curriculum, including what to teach, how and when to teach it, and how to assess changes that have occurred as a result of the instructional experiences.

- By providing time and scheduling flexibility, administrators actively support implementation of what teachers have learned: new strategies, curriculum development, research, reflection, collegial cooperation and sharing, and leadership.

- Because they are participants and understand the process, administrators become advocates for the teachers, supporting the changes being made with district personnel, school board members, parents, and the public at large.

- Collegiality, collaboration, and sharing of ideas and experiences are integral components.

21

- Practice, feedback, and time for reflection are part of the experience.

- Teachers are encouraged to become leaders.

Is it because we are not paying attention to what is really happening in the students in regard to the concepts we are trying to help them develop?

Publishers and school districts spend hundreds of thousands of dollars on educational materials for teachers to use in classrooms. Since it is an expensive and potentially lucrative enterprise, thousands of developmental hours go into the production of these materials, including consultation with classroom teachers, university educators, and state departments of education. With such potent resources, why aren't students learning more effectively? Why do they not look at the world differently as a result of their classroom experiences?

Part of the answer lies in our failure to base the design of educational experiences on what we now know about:

- How humans learn

- Diverse modalities among learners

- The role of prior knowledge in learning

Most teachers rely heavily on district curricula and commercial learning materials. Some **common flaws in curricula and learning materials**, however, are that they:

- Are too comprehensive and encyclopedic.

- Emphasize disconnected ideas and terminology.

- Present abstractions that are beyond the developmental ability of many learners at a given grade level.

- Are poorly written.

- Present incomprehensibly truncated explanations.

- Contain misleading or erroneous presentations of concepts.

- Overly emphasize reading, memorization, rote tasks, and "right-answer" assessment.

- Assume that activities are to verify written information rather than to facilitate learning.

A growing body of research on how students learn science centers on a constructivist view of learning and the role played by the preconceptions learners bring into the classroom. These previously formed ideas also are called naive ideas or alternative views (Chapters 3 and 6). Often they are incomplete understandings and sometimes they are outright misconceptions. Because the early work in this area focused on misconceptions, the whole area of study is sometimes called "misconceptions research," although this broad application of the term is not strictly accurate.

As explained in greater detail in Chapter 6, these pre-existing ideas sometimes stand in the way of students learning the desired explanations for natural phenomena. They are innate conceptual blocks that even the learners do not recognize as such and that do not yield without being specifically identified and challenged. Typically, however, learners are required to simply *accept* the version that is offered. That approach assumes that they can learn it from the experiences provided and the textbook presentation. **Seldom is the instructional process interrupted if they "get it wrong" to determine why learning has not occurred or to identify the nature of the concepts that have been formed.**

A consistent focus of the instructional model practiced in the WyTRIAD is on finding out **what** learners believe in regard to a concept and **why** they hold those views.

- Investigation focuses on what the learner brings to the classroom in regard to a concept before designing instruction about the concept.

- Only knowing both the starting point and ending point permits determination if change has occurred and what knowledge (meaning) has really been constructed.

- Only by studying students' responses to a concept is it possible to determine if the concept and the way we attempt to teach it are appropriate for the learners in our classrooms.

- The information that comes from listening to what students say and observing their responses to instructional activities and materials is the <u>data on which to base decisions about curriculum.</u>

Changing views about teaching and learning

If we accept that, in the final analysis, it is the teacher who can change the classroom, the teacher needs to be fully equipped and fully empowered to create change. In order for a teacher to be an instrument of constructive, informed change, she/he has to have an up-to-date, functional understanding of the teaching-and-learning process.

The WyTRIAD provides the teacher with the opportunity to experience and implement the constructivist learning philosophy through an integrated repertoire of research-based teaching strategies. This extended opportunity involves:

- Interviewing students

- Implementing effective teaching strategies

- Observing the changes in students;

- Examining the appropriateness of the curriculum, textbooks, other instructional materials, and assessment strategies

- Interacting and discussing with colleagues, administrators, and the professional development partner

- Examining his/her own rationale for the teaching-and-learning process

- Making appropriate changes

Some of the characteristics of the teaching-and-learning process as it is presently conducted are:

- Curriculum, instruction, and assessment are driven by behavioral objectives that are narrow in scope, decided largely by district guidelines and textbooks.

- Instruction is very direct and teacher-centered, usually emphasizing lectures, demonstrations, procedures, words, formulas, confirmation or verification activities, reading, exposition, and limited (if any) "hands-on" activities and genuine investigations.

- Teachers focus on instructional techniques, classroom management, and a behavioral-objective driven curriculum.

- Most of what happens in the classroom is dictated or demanded by others who are outside the classroom and the teacher makes her/his best attempt to interpret and implement these edicts.

- Standardized tests given at prescribed grade levels are a strong driving force for what happens in the classroom, putting teachers under pressure to "teach to the tests."

- Present assessment heavily emphasizes the cognitive domain, especially vocabulary; *e.g.,* pencil-paper, true-or-false, multiple-choice, matching, and fill-in-the-blank performance items that focus almost entirely on information acquired by memorization and that measure the reading ability of the student rather than conceptual development.

- Assessment is geared to assigning grades, rather than to providing feedback and encouraging mastery.

Change in the teaching-and-learning process requires examining the rationale and effectiveness of present practices, as well as proposing and testing alternatives. There should be good reasons for doing what we do. Asking a variety of questions about what we are presently doing gives us a fresh perspective on it. Some of the kinds of questions to ask are:

- Are we demonstrating "best practice" based on the research on teaching and learning?

- Do both our expectations of students and the activities and other classroom experiences we provide reflect current national standards and guidelines?

- Does our instruction take into account a constructivist view of learning?

- How do we know the learning experiences we are providing for students are really effective?

- Does our assessment philosophy and practice include a variety of strategies, including performance, interviews, observations, projects, and other genuine means of measuring student growth and development, habits, skills, learning, conceptual understanding, and attitudes?

Individual teachers also have specific kinds of questions to ask as they make instructional decisions, such as:

- Is what's included in the text and other instructional materials appropriate for the children in my class?

- Should I try to cover all the material that is presented in the book?

- How do I make a decision about what not to use or what to change?

- Should I do all the activities?

- Should the kids answer all the questions?

- How do students' ideas relate to the ideas as they are presented in the text and other instructional materials?

- What concepts do *Benchmarks* and the national standards consider to be appropriate for students at the grade level I'm teaching?

- How can I effectively integrate an idea or activity I learned about from an in-service workshop, magazine, meeting, or colleague?

As you can see, we are among those who believe that it is important to examine the teaching-and-learning process. Many present practices have serious shortcomings, and through asking questions and problem-solving, they can be identified and improved. An important point of the WyTRIAD is to use the information that is being collected as a result of this questioning to formulate a rationale for what to do in the future. In addition, we want to demonstrate why we should retain some practices and why should we discard those that we feel are not effective. Without a rationale based on **investigation and evaluation** we cannot generate the criteria for making decisions to change.

Finally, no individual teacher, administrator, or educational researcher can claim to possess all the knowledge and skill that is useful or important for making changes in the teaching-and-learning process. A cadre of partners is needed to create a meaningful change. Best practice doesn't happen in isolation. Diversity and collaboration provide a variety of perspectives to help make the best decisions, not only in developing criteria, but also in prioritizing and organizing their implementation

Chapter 2

How can problems
be addressed?

New and higher standards for what all students should know and be able to do; new state and local curriculum frameworks and alternative student assessments aligned with the voluntary standards; rapidly expanding knowledge, new technology, and an ever-increasing diversity of students to serve: This is education today. ...an essential dimension in responding to these new demands is high-quality, career-long professional development. It is important not just for teachers, but also for administrators, curriculum specialists, teacher educators, paraprofessionals, school board members, and many others. They all contribute to teaching and learning environments where professional development can be transformed into improved instruction and increased student learning.

*... Office of Educational Research and Improvement
United States Department of Education*

What are the problems?

Our society has changed drastically since the days when free public education based on democratic, utilitarian ideals was instituted. Today's students and teachers deal, on a day-to-day basis, with a complex, confusing, and time-consuming array of social, political, legal, administrative, and health-and-safety-related factors that make it impossible for students, teachers, and administrators to focus "just" on teaching and learning.

At the same time, we are called as a nation to respond to international educational comparisons--particularly in science and mathematics--and the increasingly sophisticated needs of employers to maintain international economic competitiveness. Clearly, all parties involved in the education of children must themselves be active, lifelong learners, as the opening quote suggests.

Our national dialogue on the need for improvement in education is focused on several key points, fueled by the comparatively low performance of U. S. students on internationally administered examinations. Some of these problems are summarized below, as they relate specifically to the rationale for teacher education, both pre-service and in-service.

- **Lack of attention to certain disciplines, such as science**

Science has not been given adequate attention, particularly at the elementary level. Many elementary schools across the country report teaching science less than one hour a week.

- **Low student achievement in science and mathematics**

A recent international study reveals that U. S. 13-year-olds are at or near the bottom in mathematics and science achievement compared to a number of other countries, city-states, and some Canadian provinces.

- **Decline in student attitudes**

There is a major concern among teachers and other educators that many elementary school children have developed a negative attitude toward several school subjects--particularly the sciences--by the time they reach third grade.

Why are there problems?

A list of some of the possible factors contributing to these problems is presented in Figure 2.1. Several of these factors (below) are of special concern in this book.

- **Mismatch of curriculum, materials, and classroom strategies to existing intellectual level of the learner**

Particularly in science, many topics taught in school are inappropriate for the developmental level of the intended learner. For example, Epstein (1979) reports that over 80% of 13-year-olds are concrete thinkers with respect to many concepts. In spite of such studies, however, most concepts taught to students at this age <u>and</u> the books used to teach them require much abstract thinking.

- **Lack of relevance**

There appears to be a lack of connection between what is presented in the classroom and what goes on in children's lives. As a result, the concepts taught in school often are not integrated into children's explanations about phenomena they observe and experience.

- **Curricular decisions not appropriately based**

In many cases it is not the classroom teacher, or even the author, who determines what should be in the curriculum and published materials. Publishers include things that will sell, based on feedback from state departments of education and district adoption committees. The result is perpetuation of traditional topical presentation, in spite of the experiences of teachers and research that demonstrates inappropriateness of level, sequence, or manner of presentation.

- **New ideas (trends and research) not usable by practitioners**

Much of the educational research is either inaccessible to the classroom teacher or is communicated in a manner that is intended for other researchers. Classroom teachers are kept too busy to have time to study, reflect upon, adapt, and assimilate educational research, trends, and recommendations.

- **Low professional image of the classroom teacher**

Unlike many other civilized nations, America does not hold its classroom teacher in high esteem. The classroom teacher is one professional who is constantly challenged by just about everyone who has gone to school. Teachers are among the lowest paid professionals, are often treated as semi-skilled labor, and are given few incentives and little support for professional growth and improvement.

Why are there problems?

- Mismatch of curriculum and materials to the learner

    » Expectations of students not appropriate to their developmental levels and prior experiences

    » Rigidity of the curriculum

    » Testing, assessment, and grading policies

    » Emphasis on teaching rather than on learning

    » Assembly-line philosophy that applies the same approach to all learners

- Lack of relevance and connection to students' lives

- Inappropriate curricular decision-making process

- New ideas not in a form that is usable by practitioner

- Less-than-professional image of the teacher

- The "mysterious, beyond-comprehension" aura of the subject matter

- The attitude that learning mathematics and science requires special ability, which most students do not have

- School structure and assigned roles and expectations

- Parental attitudes and lack of community understanding of the teaching-and-learning process

Figure 2.1 Some possible causes contributing to student difficulties with mathematics and science.

Vision, rationale, and needs underlying the WyTRIAD model

As noted in the Prologue, the ideas that are incorporated into the WyTRIAD in-service model support the philosophy that, since teachers are the ones who are charged with implementing the curriculum and meeting the needs of students, they also are the primary agents of change in a school. Conversely, they can be strong forces for maintaining the *status quo*. For meaningful changes to take place in the classroom, the teacher must both value and recognize the need for change and have the opportunity, support, and skills to create desired changes.

The **vision** directing WyTRIAD is that the teacher, as a professional, can take control of the teaching and learning process by:

> • Employing a philosophy of education grounded in our present, research-based understanding of teaching and learning
>
> • Becoming a researcher of the teaching-and-learning process
>
> • Using classroom-based research to design, make decisions about, and implement meaningful curriculum--that is, what to teach, how to teach it, and how to assess the nature of the learning that has occurred

Research evidence and the experience of teachers show that significant change only occurs *in* the classroom ("where the rubber meets the road"); yet, for decades our educational system has followed a direct, linear model of decision-making, starting with national, state, or local administrators. Teachers have typically been--or at least have perceived themselves as--the recipients of "top-down" policies, procedures, and curricula (Figure 2.2).

Often, attempts to involve teachers in the decision-making process have the teachers' ideas and participation essentially inserted into the process from the side (Figure 2.3). This type of participation still (1) does not change the direction of the decision-making process, nor does it address (2) the needs of the educational system to take into account advances in our knowledge of teaching and learning or (3) the needs of teachers to have opportunities to learn, grow, see

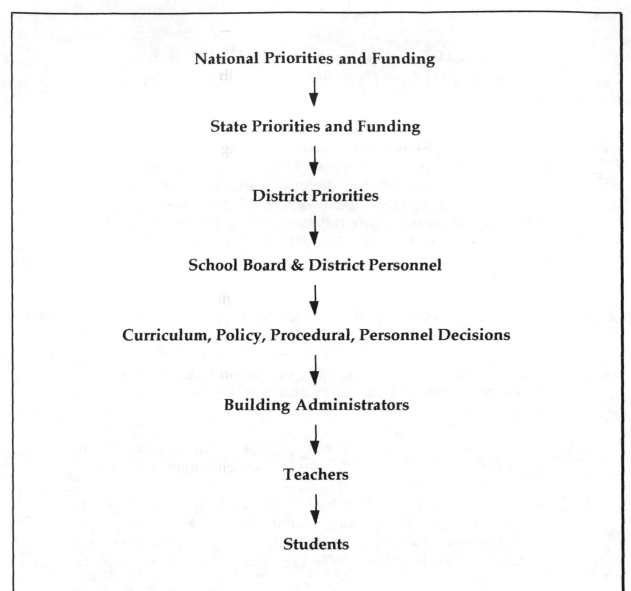

National Priorities and Funding

↓

State Priorities and Funding

↓

District Priorities

↓

School Board & District Personnel

↓

Curriculum, Policy, Procedural, Personnel Decisions

↓

Building Administrators

↓

Teachers

↓

Students

Figure 2.2. A typical linear model of educational decision-making and implementation.

modeled and practice new strategies, develop collegial teams, and become reflective practitioners.

A major difference between the WyTRIAD model and most of the other ways teachers are involved in educational change is that **teachers are empowered to make informed decisions based on evidence they collect in their own classrooms.** The teachers become the knowledgeable experts on what is happening (1) with the students in their classrooms in their communities, and (2) with their own professional development. They become empowered to implement informed decisions and to provide leadership in directing appropriate educational change.

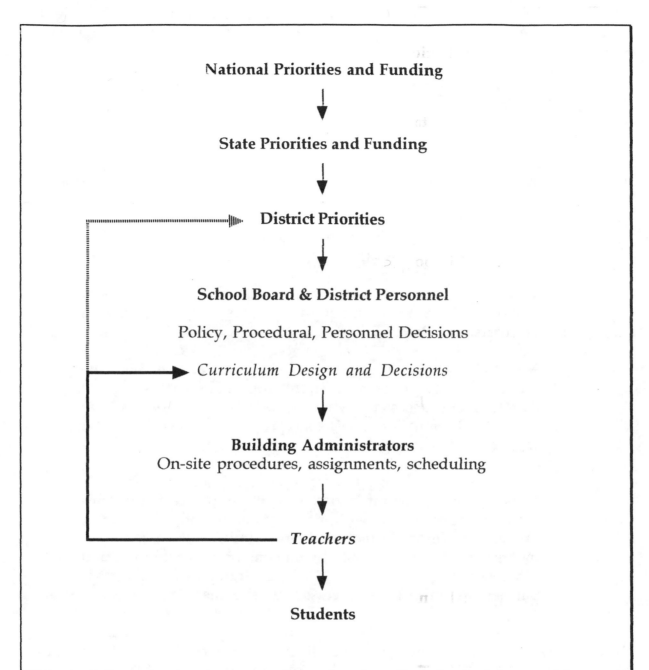

National Priorities and Funding

↓

State Priorities and Funding

↓

District Priorities

↓

School Board & District Personnel

Policy, Procedural, Personnel Decisions

Curriculum Design and Decisions

↓

Building Administrators
On-site procedures, assignments, scheduling

↓

Teachers

↓

Students

Figure 2.3.   A common model of educational decision-making and implementation in which teachers are invited to provide their ideas on curriculum and district priorities that are related to curriculum.

A further distinction of the model is that it is **not a prescriptive, rigid model** that is characterized by imposed priorities, behavioral objectives, checklists, worksheets, lesson planning forms, or a universal, static curriculum. It **is** a flexible, dynamic model that is based on all participants **asking questions** about the needs of their own students, schools, and themselves as professionals, then making **research-based decisions** about designing appropriate teaching-and-learning experiences, selecting curriculum materials, and otherwise restructuring the school environment.

Teachers cannot accomplish change alone, however, nor should they be expected to do so. It takes a **partnership**. This relationship is the critical element missing in most other models and operating systems for developing curriculum and making decisions about what will transpire in the classroom. In the WyTRIAD, there are three essential partners--teacher, administrator, and outside resource person. Each partner brings **special knowledge and expertise** to the relationship.

> • The <u>teacher</u> has immediate and daily contact with children and, through the strategies introduced and practiced in the WyTRIAD, learns how to document and create change. The teacher becomes the decision-maker in the classroom environment rather than simply implementing the decisions of others.
>
> • The <u>administrator</u> establishes and administers the working environment, including scheduling, staff development, assessment, etc. The administrator has the power and ability to provide support and opportunities for the teacher to learn, grow, and create change.
>
> • The <u>professional development facilitator</u> provides assistance and current knowledge about the teaching-and-learning process; national trends, standards, and frameworks; newer and more effective instructional and assessment models and strategies; subject area content and activities; and other matters. The ability to model strategies for the other partners is an absolutely necessary part of this expertise.

The relationship that is created becomes an interdependent configuration in which the opportunity and expertise all three partners are essential (Figure 2.4). If any one piece of the partnership is not fully committed, the relationship is broken and the process is again fragmented and linear. **A central feature of the model is that it is completely organized around what is actually experienced by students in the classroom.** Information (data) about (1) what students bring to the learning situation and (2) how they respond to the experiences guided by the teacher are the basis for making decisions about curriculum, instruction, and assessment.

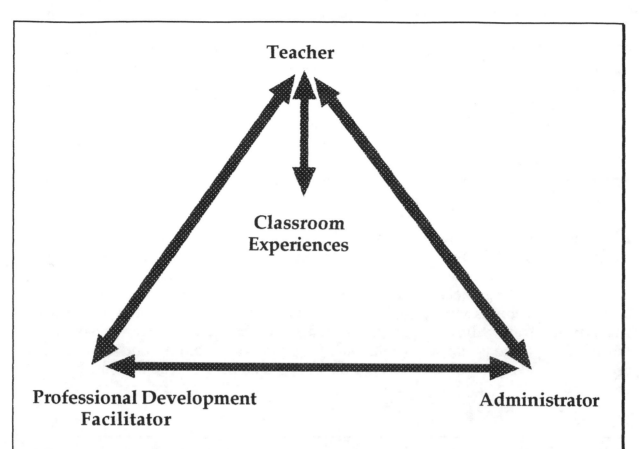

Teacher

Classroom Experiences

Professional Development Facilitator

Administrator

Figure 2.4. Interlocking, interdependent pattern of assistance, advice, support, and decision-making in the WyTRIAD model, characterized by cooperation, collaboration, and shared expertise. The area enclosed by the triangle represents all of the factors that come together when a student walks into the classroom, including prior knowledge, curriculum, learning resources, the teacher, peers, and home.

WyTRIAD origin

The WyTRIAD program was developed as an attempt to respond to the apparent inability of research-based knowledge to be effectively implemented in the classroom. It is described as "a partnership for changing the school from within" (Stepans, 1989, 1990, 1993). The word "TRIAD" was used because the model is structured as a three-way partnership of classroom teachers, their administrator, and a professional development facilitator. The generic term "triad" has been applied to other types of partnerships (home-school-community, other professional development models); however, the Wyoming TRIAD in-service experience is a unique combination of integrated components.

The first WyTRIAD began in 1990 when letters of inquiry were sent to the superintendents of all 49 school districts in Wyoming, inviting them to consider participation in what was first simply called the TRIAD project. Of the 49 districts, three responded. Arrangements were made to make a presentation to school administrators and interested teachers. The presentation consisted of reviewing the importance of research in education and the role of teachers as leaders in educational research. The importance of the partnership and the active roles of each partner were explained. A tentative agenda, the roles and responsibilities of the partners, and the school commitments were outlined.

Goshen County was the first school district to accept the challenge and became the first TRIAD site in Wyoming. Twenty-seven K-9 teachers representing five area schools, two building administrators, and an assistant superintendent participated in the project. There has been a great difference between the success of TRIAD in schools where the building administrator participated and where the administrator did not get involved. Goshen county now has been participating three years, and a number of districts in Wyoming and other states are either actively involved in WyTRIAD programs or are preparing to become involved.

What are the goals of the process?

Altering teachers' fundamental perceptions of the teaching and learning process, changing what and how teachers teach, changing how they assess, and transforming teachers into applied researchers are explicitly listed as goals of the WyTRIAD (Figure 2.5). The WyTRIAD model seeks to provide a cost-effective, **site-specific** in-service experience with a high likelihood of subsequent implementation--one that will persist and help to create real change.

It takes advantage of the requirement for in-service training for all teachers, in that teacher participants usually enroll in a class taught by university personnel. The support of the administrators must be in evidence before the class is begun and administrators' attendance at the class sessions is highly encouraged--it is stated as a prerequisite, with the hopes that administrators will honor the commitment.

Often, the WyTRIAD is offered as a graduate course through a university partner. There are several key advantages to this being a graduate-credit arrangement. It sets a specific learning modality. There is a motivation to give

The WyTRIAD aims to transform the classroom by providing teachers and administrators with the opportunity to:

1.   Become aware of the research on teaching and learning.

2.   Become aware of science and mathematics misconceptions in themselves and their colleagues, as well as those that could be created by their classroom materials (such as textbooks, workbooks, curriculum guides, and models).

3.   Identify student preconceptions and misconceptions and use this information in curriculum design to help students overcome their misconceptions and raise their conceptual level of understanding.

4.   Apply research knowledge about how children learn.

5.   Learn how to teach and assess for conceptual change, explicitly taking into account what students bring to the classroom based on their prior experiences.

6.   Participate in peer coaching.

7.   Conduct classroom research and apply the results to their own teaching.

8.   Make critical, professional decisions about what constitutes appropriate content, strategy, and assessment for teaching specific science and mathematics concepts--a skill that is transferable to other subject areas.

9.   Be more reflective in their approach to teaching and to use what they learn to improve the teaching-and-learning process in their classrooms.

10.   Develop strong collaborative teams in schools, with teachers, administrators, and professional development persons working together to improve the learning experiences of children.

Figure 2.5. Goals of the WyTRIAD in-service model.

the activity a high priority in already busy lives. It enhances commitment from and persistence of <u>all</u> partners. Furthermore, unlike many in-services, the class is ongoing over a four-to-five-month period.

Five class **sessions** range in length between three hours and two full days. The **intervals** between the class sessions, three or four weeks, are used for preparation and implementation of course ideas. During the sessions, university personnel (or others serving as facilitator) meet and work with the teachers and administrators. During the intervals, teachers implement the ideas, strategies, and activities in their own classrooms. They collect data and maintain reflective journals. Class sessions are rich with interplay among all partners, discussing what they have done, what they have observed, and what they plan to do next.

Critical to the WyTRIAD model is the treatment of **teachers** as equal educational partners and people from whom the others can learn. One of the important roles they play is that of researchers. If teachers do the research rather than just read about it, there is a greater chance that they will implement the ideas (Butzow and Gabel, 1986). The teachers also determine the conceptual content around which the course is built, implement the teaching strategies discussed, and make judgments about the appropriateness of course content and curriculum to the learners. Because they are involved in many innovative activities while participating in WyTRIAD, teachers must become comfortable risk-takers if they are not already.

The involvement of the **administrator** is the second piece of the three-way partnership. Since, by definition, the administrator actively participates in WyTRIAD, philosophical and financial support for the teachers is built in. At the same time, the administrator comes to understand the needs and concerns of the teachers as professionals. The teacher-administrator relationship under the WyTRIAD model is designed so that cooperation is fostered.

The third piece of the WyTRIAD partnership is the **professional development facilitator**, usually a university professor, although it also may be school district personnel and/or a lead teacher or teacher team with WyTRIAD experience. The primary purpose of this partner is to provide expertise, feedback, and resources for the other two partners. He or she introduces and models the instructional strategies and interviewing techniques and provides, with the aid of the administrator, peer coaching opportunities. The professional development person assists the classroom teachers in using what they have learned through the other components of the class to make meaningful, practical changes in what and how they teach and how they assess. In each of these roles, the professional development person serves as a guide rather than as a sage. While he or she may have more incoming research knowledge, the WyTRIAD's working viewpoint is that change that results from teachers' own classroom research is far deeper than change prescribed from outside.

Helpful resources in this book

This book is designed to be used by all three partners during a WyTRIAD experience, <u>as well as</u> to be useful to all educational professionals who are interested in **practical, integrated, research-based teaching strategies and in-service models**. Since our vision is to create a classroom environment that is conducive to learning and where effective and meaningful changes are taking place, the book presents a blend of theory and practice in which the role of the teacher as an investigative decision-maker and facilitator of learning is central.

Classroom teacher

The book is a resource to the teacher to provide information and support as he/she goes through a WyTRIAD experience. It is to be used as a practical handbook, textbook, and reference through the process. It also provides information and experiences that can help teachers become decision-makers about the teaching-and-learning process, leading to genuine restructuring in the classroom. The book has:

- Background information about

  » What constitutes an effective in service model
  » Main components of the WyTRIAD in-service model

- In-depth discussion of major topics

- Review of and information about

  » Research-based perspectives on learning
  » Constructivist approach to both the teaching-and-learning process and to professional development

- Examples, materials, and forms to assist with

  » Student interviews
  » Peer coaching
  » Classroom research
  » Data collection, keeping track of observations
  » Suggestions for various types of reflection

- » Journals
- » Lesson plans, using a conceptual change model
- » Assessment
- » Aligning curriculum, instruction, assessment

- Quotes and testimonials from teachers who have experienced the model

- Descriptions of specific, practical activities, behaviors, and teaching strategies

Administrator

The active participation of the building administrator is <u>essential</u> for teachers to be successful in implementing the ideas for change that are advocated here. For administrators, the book also:

- Provides information on the change process

- Describes an effective partnership to create change, emphasizing team-building

- Identifies importance of administrator's role as a partner

- Provides evidence of the contrast in effectiveness between sites where the administrator has and has not fulfilled his/her role

- Provides a practical overview and specific ideas for implementing a constructivist philosophy in classrooms and as a part of professional growth

- Has explicit guidelines for organizing and participating in the experience, enabling advance calendar and staff planning and scheduling

In-service facilitator

The leadership role of the facilitator is critical to the development, continuity, and content base of the experience. The book provides a common base of information and understanding with the other partners, as well as making clear the role of the facilitator. Particularly, for the facilitator, the book:

- Is a guide to conduct the model, containing

    » Background information
    » Specific agenda for each session
    » Suggestions for initiating and conducting the process

- Provides information to augment sessions

    » Background information
    » Relevant topics
    » Activities
    » Forms

- Suggests an evaluation process

For all participants and readers

The background information and translation of research reocmmendations and educational theory into practical applications should be helpful to all interested readers. A selected list of **references** and helpful **appendices** are provided. In addition, the book suggests:

- How to initiate and set up a WyTRIAD experience

- How to share, disseminate, and advantageously use information gained from the experience

- Ways to influence understanding, knowledgability, and decision-making processes beyond the individual school (e.g., district level, school board, parents, other publics)

Chapter 3

How does WyTRIAD incorporate a constructivist approach?

Students' prior knowledge needs to undergo change for the new scientific ideas to fit... What makes this change problematical is, in part, the role played by prior knowledge. Learning is not simply a matter of adding new knowledge, nor a matter of correcting incorrect information. The prior knowledge includes the interpretive frameworks that the learner uses to make sense of the world and to communicate with other people. It is these interpretive frameworks that must change.

... Edward L :Smith, 1991

What is the constructivist view of learning?

The teaching for conceptual change strategy is based on a **constructivist** view of learning, the idea that each individual mentally constructs his or her own knowledge. In this view, children make sense of the world and the things in it in their own way, based on their own experiences. As a basis for instructional practice, constructivism requires us to examine both the way we view learners and the way we view ourselves and our approach to the teaching-and-learning process.

Constructivism acknowledges that important learning occurs outside the classroom as well as inside. It occurs any time and anywhere that a person is perceiving and interpreting information from her or his environment. The way we come to learn about the world and the things in it is a result of our own experiences, our interactions with our environment, other people, and materials.

Learning is a construction of meaning from life's experiences that begins in infancy, long before formal schooling. Rather than simply "filling up" with and accepting new information, our brains work to actively **fit our interpretations of new experiences into our existing framework**--what we already know. Our brains even make up "memories" to fill in gaps in understanding. The results of this fitting together are often problematic. Students typically enter the classroom with novel ideas that make sense to them and that fit together what they know, but that do not match the lessons they are about to experience. The result can be a frustrating and difficult clash between the students' personal and strongly held explanations and the beliefs and explanations that are imposed on them by the authority of the teacher and the textbook. Often, there is no change in the usefulness and acceptability of the students' own interpretations, and the students learn "for the test" rather than constructing meaning.

A constructivist approach to teaching, on the other hand, reveals and takes into account these pre-existing ideas. As a result, learning involves **conceptual change**--a change in the concepts learners bring to the classroom with them and/or a different way of fitting the new ideas into their existing frameworks.

Clements and Battista (1990) compare a constructivist view about teaching and learning to a traditional view of learning based on transmission and absorption of information. According to them, constructivism involves five components.

> • Knowledge is invented by the learner, it is not something that is passively received from the environment.
>
> • Ideas are made meaningful when learners are able to integrate them into their own existing structure of knowledge.
>
> • No one, true, reality exists, only individual perceptions of things.
>
> • Learning is a social process and a constructivist environment is a culture in which the students are involved in explanation, negotiation, sharing and evaluation.
>
> • When we demand that students memorize certain rules and procedures or accept certain explanations we curtail their sense-making activity.

What is meant by teaching for conceptual change?

Teaching for conceptual change applies the constructivist view of learning. It is geared to bringing about changes in the learner's **existing** concepts, understandings, and mental models. The strategy focuses on the concepts, knowledge, and beliefs the learner brings <u>to</u> the instructional experience, as well as what the teacher wants the learner to gain <u>from</u> the experience. It emphasizes being aware of what actually happens within the mind of the learner.

It begins by helping students to verbalize their existing views so they and the teacher share a starting point for the instructional experience. Instructional experiences are then provided that cause students to confront their existing views, testing them by doing activities. A mismatch between what a learner predicts based on a pre-existing belief, and the result of his or her own testing creates mental conflict. The desire or need to resolve the conflict motivates learning, leading to conceptual change.

Several conditions are necessary for conceptual change to occur (Posner, et al., 1982; Strike and Posner, 1985; Hewson, 1981; Hewson and Hewson, 1984).

- The student must be dissatisfied with her or his existing views.

- The new conception must appear to be plausible.

- The new conception must be more attractive than the previous view.

- The new conception must have explanatory and predictive power.

How do we aid students in constructing knowledge?

The classroom environment

Constructing new knowledge involves risk, including the risk of exposing one's misunderstanding or lack of knowledge, and the risk of letting go of personal beliefs and explanations in favor of another view. Classrooms can be, from the learners' point of view, uncomfortable and perhaps even hostile places. To facilitate the teaching-and-learning process, therefore, we need to **create an environment** that is conducive to learning, accepting of everybody's ideas, and that offers diverse opportunities to help students to make sense of things in their own ways.

To construct and become secure in their new knowledge and understanding, students need to go through several stages in regard to the new concepts they are learning. **The constructivist classroom provides opportunities for:**

- Having learners <u>become aware</u> of their own preconceptions and beliefs about a concept

- Helping students to <u>share</u> their ideas and become aware of the ideas of others

- Encouraging students to <u>test</u> their ideas by working with materials or going to the sources

- Helping students to <u>make sense</u> of the concept and <u>resolve the conflict</u> between what they thought and what they experienced during the activity

- Helping students to <u>apply</u> what they have learned by making connections to other situations

- Encouraging students to pursue <u>additional</u> questions and problems related to the concept

Conceptual change takes time. Preconceptions are typically strongly held. Some learners, as the result of a single striking experience, may be able to instantly accept the clarification of their previous ideas. Generally, however, <u>repeated experiences</u> are necessary to replace naive preconceptions and misconceptions with the scientific explanation of a concept. A **variety** of activities, in different settings and contexts, needs to be experienced.

The Teaching for Conceptual Change Model
(CCM)

The **teaching for conceptual change model** (CCM) (Figure 3.1) is an activity-centered, constructivist instructional strategy that is central to the WyTRIAD process (Stepans, 1992, 1993, 1994). It models the kind of thinking and activity processes typical of scientific inquiry. It builds upon the traditional and popular learning cycle (Atkins and Karplus, 1962), but goes beyond the learning cycle in several meaningful ways, taking into account new knowledge and perspectives in cognitive science and science education that have developed since the learning cycle was introduced more than 30 years ago.

The CCM consists of a sequence of six stages that mirror the research on conceptual change. Its components are based on the experiences, ideas, and research of many teachers, science educators, and others. Some seminal references are Atkins and Karplus (1962); Eaton, Anderson, and Smith (1983); Clement (1987); Nussbaum and Novick (1981); Posner, Strike, Hewson, and Gertzog (1982); Driver and Scanlon (1989); Duit (1987); Feher and Rice (1985); Gilbert, Osborne, and Fensham (1982); and Stepans (1987, 1988, 1991). Some of the terms used in the model were first proposed by Feher and Rice (1986), and have been adapted here.

The CCM is an inquiry learning model in which the teacher and student are both learners--the teacher is no longer the answer-holder. Both students and teachers confront change in themselves through the use of the model. The students use the steps of the model (**predicting, sharing predictions and explanations, testing, resolving the concept, building connections, and leaving the topic open for future questions**) to learn about a science concept. The teacher may use many of these same steps to gain an understanding of the children's attitudes, socialization, knowledge, and skills. One of the strengths of the model is that it enables teachers to more accurately judge the appropriateness of the curriculum for the learners in her/his classroom.

One of the most striking **outcomes** of this strategy that is reported by teachers is that many students who have difficulty with traditional, book-based instruction do well using the CCM. Also, the teachers' observations help them to look at kids differently, to acknowledge and value the ideas learners already have, and to build upon them. Furthermore, through active collaboration, children learn to value and respect each other's ideas.

The Teaching for Conceptual Change Model (CCM)

1.   Commit to an outcome

Learners become aware of their own perceptions about a concept by thinking about it and *making and explaining their reasons for predictions*--committing to an outcome--before any activity begins.

2.   Expose beliefs

Learners expose their beliefs by *sharing predictions and explanations*, initially in small groups and then with the entire class.

3.   Confront beliefs

Learners confront their beliefs by *testing and discussing* in small groups what they observe from doing the activities and collecting data.

4.   Accommodate the concept

Learners work to accommodate the concept by *resolving conflicts* (if any) between their initial ideas (based on the revealed preconceptions and class discussion) and their observations.

5.   Extend the concept

Learners extend the concept by *trying to make connections* between what they have learned in class and other situations, including daily life.

6.   Go beyond

Learners are encouraged to go beyond by *pursuing additional questions and problems* of their choice related to the concept.

Figure 3.1. This Conceptual Change Model (CCM) for teaching and learning aids learners in constructing their own knowledge by causing them to explicitly acknowledge and challenge existing understandings (preconceptions) through concrete experiences (Stepans, 1992, 1993, 1994). Further, it asks the learners to make connections, apply the target concept, and propose additional ideas and tests.

How do teachers become constructivists of the teaching-and-learning process?

In the WyTRIAD, students are not the only learners. The process also encourages the teachers to become **constructivists of the teaching-and-learning process**. By trying new strategies and collecting information in their classrooms, they are constructing their own knowledge about what works or doesn't work and why. Once teachers understand how the kids construct meaning through diverse experiences, they come to see themselves and their professional activities that way too. Thus, **two** types of constructivism are going on in WyTRIAD. The <u>students</u> are engaged in active learning about their course concepts. The <u>teachers</u> also are trying to make sense of the teaching-and-learning process.

Several specific components of the in-service model help teachers become constructivists of the teaching-and-learning process. They engage in a parallel to the process of **prediction and testing** experienced by the students. Before they begin teaching a lesson or trying a new strategy, teachers make predictions about what *they* think will be the students' behavior during the lesson. These predictions could also be called "anticipated outcomes." A format such as that shown in Figure 3.2 suggests a method for **documenting** change within the teacher who adopts a constructivist approach to his/her own learning. The same type of approach can be used in anticipation of an interviewing, peer coaching, sharing, or planning experience, or a meeting with parents or an administrator. The process also could be extended to the administrator and professional development partner, encouraging them to maintain documentation and reflect on what they see and experience.

This process of **expectation, observation, and reflection** is an integral part of each WyTRIAD activity. Participants keep a journal, which is intended to be a working companion in which they record their ideas and predictions, data (observations), and subsequent thoughts about what they are experiencing and what their observations mean.

As the teachers do interviews, implement new teaching strategies and experiences, make observations, collect data, and make inferences from their data, they **construct their own philosophical frameworks** about the teaching-and-learning process. This framework is based on what they already know, what they learn from the professional development facilitator, their colleagues, their administrators, and--most significantly--what they learn from their own direct experiences. This framework consists of <u>actively constructed</u> knowledge and becomes the theoretical basis for their professional practice.

The process of **consciously** attending to personal and professional change and growth is what we mean when we say that one of the goals of the WyTRIAD is to help teachers become constructivists of the teaching-and-learning process.

Class: Today's date:

A. Expectations: (My predictions regarding what will be the students' behavior,
                               attitudes, etc., during the lesson)

   1.

   2.

   3.

B. Observations: (What I observed)

   1.

   2.

   3.

C. Reflections: (How I resolved any difference between what I predicted and
                               what I observed. Other ideas.)

   1.

   2.

   3.

Figure 3.2. Sample format for a journal entry demonstrating a constructivist approach to teaching. The <u>predictions</u> (A) would be established before the lesson, taking as much room as needed. The <u>observations</u> (B) could be jotted into the journal spontaneously or immediately after the class (as soon as possible). <u>Reflections</u> (C) would be entered whenever the ideas were synthesized. The numbering system, keyed to individual predictions for the day, is for easy reference between sections of the entry, which may take more than one page.

Chapter 4

What is meaningful restructuring of the teaching-and-learning process?

The vision of science education described by the Standards will require changes throughout the entire system. Although teachers are central to the education of students, they alone are not responsible for reform. Teachers must be provided with resources and time, and they must work within a collegial, organizational, and policy framework that supports reform efforts. The vision described by these teaching standards will not be attainable without major changes in the educational system.

... National Science Education Standards
National Research Council, 1995

What is meant by "restructuring?"

Restructuring is a popular word right now. In this book, we are not necessarily talking about physical remodeling--knocking down walls or adding cabling to accommodate technology. We are talking about altering the interactions between teachers and students and among students, as well as changes in the nature of the decision-making process. At its best, school structure serves the educational needs of students through making the best of all of the skills, ideas, energy, and talents of the professionals involved in the educational enterprise at a school.

Schools also serve society, and are accountable to parents, school boards, district offices, community values and priorities, and state and federal governments. With so many moving parts, a hierarchy of responsibility has developed that tends to place teachers--the professionals most in touch with students--at the bottom in terms of advice and decision-making. At its worst, school structure can be rigid, stifling, and highly political.

> *Individuals or groups dole out sanctions and rewards or assert pressures of one sort or another that disenfranchise, silence, and unnecessarily limit the opportunities of others. The ability of groups to assert pressures and the likely responses of others to them are influenced in dramatic ways by school structures.* (Gitlin and Price, 1992)

Teacher in-service is one avenue for restructuring schools, by changing the culture of the school from one that is rigid, authoritarian, and fragmented to one that is a flexible, responsive, community of learners in which teachers are valued as experts and decision-makers. One aspect of this altered culture is what is popularly called **teacher empowerment**, wherein teachers:

- Feel their knowledge and experience are valued

- Have a significant voice in school governance and curriculum decisions

- Have the freedom to investigate, innovate, and implement changes in their classrooms

- Receive encouragement and support for ongoing professional growth

Restructuring is not the individual responsibility of teachers, principals, school boards, universities and colleges, or state education departments. Rather, it is a collective challenge, requiring concerted, cooperative action.

- Universities need to provide educational experiences for teachers and administrators that are founded and grounded in an understanding of how learning occurs and in issues related to specific subject areas. In the case of science and mathematics education, for example, ideas and issues such as those raised in the American Association for the Advancement of Science (AAAS) *Project 2061: Science for All Americans*, AAAS *Benchmarks for Science Literacy*, NCTM *National Council of Teachers of Mathematics Standards*, and the National Research Council (NRC) *Everybody Counts* and *National Science Education Standards* are important.

- Teachers have a responsibility to increase their understanding of the teaching and learning process and to apply this knowledge to their students' learning experiences.

> • Administrators have a responsibility to seek and accept the expertise of others and move away from rigid management and curricular approaches that meet the needs of only a few learners.

What is needed to create meaningful change?

If there is one thing the natural world teaches us, it is that change happens. Sometimes we have an opportunity to control or direct change, but lacking such a deliberate effort, it occurs spontaneously and not always in a way that might be what we would have chosen. A backyard garden provides a useful metaphor for these statements. Things will grow there, whether they are vegetables or weeds; also, things will perish there, due to our conscious efforts to kill or remove them or due to our lack of attention or expertise in tending them.

What about change in schools? Lacking a coordinated organizational **plan** and the concerted **action** of all stakeholders, scattered and unpredictable change may occur. Individual teachers may implement new activities and strategies, others may become more deeply entrenched in old practice and philosophy. New curricula, textbooks, and management systems may be adopted, with or without an informed analysis of the impact they will have on the community within the school. Old resources may be clung to because of their familiarity, tradition, or because they were least offensive to the majority of members on some committee. University and staff development persons may continue to conduct half-day, self-contained staff development workshops, without inquiring of the recipients what would be useful to them.

There also needs to be a better <u>match</u> or <u>fit</u> between the needs or problems of the situation and the change that is made. For example, having someone come in and do a one-day workshop about science process skills will not solve the need teachers have to integrate and adapt activities in their classes to foster development of process skills. The quickie workshop may show, on paper, that *something* has been done. It is not likely, however, that true change and improvement of practice have resulted, especially if there is no mechanism for teachers to practice and evaluate the impact of the workshop on their students.

In each of the examples cited above, the missing ingredient is **data**--there is lack of an organized body of information upon which to develop action plans. In what has become known as "The Information Age" it is increasingly necessary for us to rely on knowledge developed by others. There is just too much information for each of us to manage. This same dilemma holds true for schools. Schools are more complex than ever, socially, culturally, academically, politically, financially, legally, medically, and technologically. It follows that a model for successful schools must take into account this complexity <u>and</u> the reality that partnerships are essential for dealing with both the challenges and opportunities this complexity offers. **Change cannot be prevented, but it can be guided.**

Also, change takes **time**. When people are involved in altering how they do things, they must have time to internalize, practice, implement, test, analyze, and evaluate. They can't just "get information" but need to "develop understanding," making the new knowledge fit the situation. Another way to state it is allowing a person to "develop ownership" of the concept. Just as it takes more than one exposure for a student to develop a meaningful concept, it takes more than just hearing about it or seeing a demonstration for an educator to be able to put into practice new information, a new philosophy, teaching strategy, curriculum, or means of assessment.

What are the new roles
for teachers? leaders? administrators?
staff development and university personnel?
students? parents?

Restructuring requires **changing roles**. The WyTRIAD model incorporates shifting from a linear model to one that represents contiguous and continuously interconnected roles. A shift in perspective and emphasis suggests new roles for all of the parties involved.

- **Teachers** take on more leadership and become researchers of the teaching-and-learning process.

- **University professors and other professional development persons** expand their role from instructors to learners, learning from the practical experiences of teachers and learning from the teachers what they need in order to provide useful assistance.

- **Administrators** move away from being the persons who

   » tell people what to do,
   » decide which curricula to implement, and
   » direct the events and the sequence;

to becoming the persons who

> » facilitate teachers' initiative, growth, direction, expertise, needs, and role as a source for directives,
> » make possible those things that are identified by teachers as needs to be met, and
> » finds ways to meet the needs of teachers, implementing their requests for time, materials, changes, flexibility, and freedom.

• **Teachers, administrators, university educators, and staff development personnel** work together effectively as partners to construct appropriate curriculum and practice in the classroom.

• **Students** become resources in their own education, as an interactive source of information on which to base curriculum and as active learners engaged in the construction of knowledge.

• **Parents** become active partners in the educational process.

The new roles are interdependent, and involve understanding, negotiation, and clear expectations. In order for teachers to want to implement meaningful changes in their classes, use new ideas, and take risks in changing their expectations of their students, they need the support and the understanding of their building administrators.

In order for the administrators to understand what is going on and provide support for the teachers, they need to take on a new role, one that is significantly different from the one usually expected of them. That is why they are required to attend the WyTRIAD sessions with their teachers, get involved in the activities, listen to the concerns, frustrations, ideas, and successes of their teachers, and see the impact on students. Only the building administrator can provide the teacher with the time needed to interview students, be involved in peer coaching, reflect on practice, and rethink and revise curriculum.

With the administrator fully aware and supportive of the teachers' activities, parents and district school personnel are informed and brought into the process. There is an opportunity for the teachers to participate fully in district curriculum decisions, with data to support their ideas. Some successful teachers become mentors, leaders, and workshop facilitators, make presentations at professional meetings, and publish.

How do classroom activities and materials need to change?

Classroom activities also must change as a part of restructuring. Often, this is the point most readily agreed upon by all parties, and it can be the starting point for change in the classrooms. States and districts regularly "redo the curriculum" in response to calls for reform. For the past three decades there has been an emphasis on changing the nature of the activities that are provided for learners in the classroom, and many teachers have been involved in revisions that have invoked terms such as "inquiry," "hands-on," "exploration," "investigative," as well as specific strategies such as the learning cycle and integration of science-technology-society experiences.

There is something positive and useful about all of these things. To be truly meaningful, however, changes in classroom activities require a foundation in what we are continuing to learn about the teaching-and-learning process. Just "doing" to satisfy directives or even to follow good advice may result in producing activities that are well-intentioned but misdirected and/or inappropriate and ineffective in practice.

A comparison of traditional and recommended **approaches to science education** in Figure 4.1 illustrates the challenges we face. How do we translate our current understanding of teaching and learning into effective curriculum design? Restructuring can incorporate ideas from Figure 4.1. We can use them to define **expectations** of students, bringing in ideas from the national documents about including all domains, for instance. Then, if we are to change our expectations, there should be corresponding changes in the **experiences** we provide for our students and in our **assessment** strategies. Practical alignment of expectations, curriculum, instruction, and assessment leads to new roles for the learner, teacher, and administrator. The synergistic combination of activities in the WyTRIAD model assists teachers and administrators to develop these new relationships and strategies.

Not only classroom philosophy and activities need to change and progress, so do science materials. Figure 4.2 illustrates some **characteristics of effective science materials**. The parallels between Figures 4.1 and 4.2 demonstrate alignment of theory and application.

Changing Approaches to Science Education

Traditional

- Science for some

- Behaviorist-based

- Behavioral objectives: learning based on measurable objectives

- Text-based

- Passive

- Confirmatory investigations

- Science seen as a single subject

- Teacher imparts knowledge, student learns it

- Limited use of technology

- Competitive learning

- Exclusive use of paper and pencil assessment

- Many science topics covered, with little depth

- Single exposure

Recommended

- Science for all

- Constructivist-based

- Conceptual objectives: learning based on constructing meaningful concepts

- Hands-on and minds-on

- Active

- Problem-solving investigations

- Science seen as a part of an interdisciplinary world; emphasis on relating science to students' world

- Teacher as a facilitator of learning, student as learner who constructs knowledge

- Full integration of appropriate technology

- Cooperative/collaborative learning

- Multidimensional assessment; assessment integrated with instruction

- Fewer science topics covered, with more depth

- Spiral curriculum

Figure 4.1. A comparison of traditional and recommended approaches to science education. Adapted from Wright, E., and J. Perna. (1992). Reaching for excellence: a template for biology instruction. Science and Children 30(2):35.

Effective science materials

- Are centered around hands-on investigations of scientific phenomena

- Are developmentally appropriate for the intended level

- Stimulate student inquiry

- Interest students

- Have potential for exploration

- Explore important science topics in depth

- Teach science process skills in the context of investigations

- Teach students to work together in teams

- Are appropriate for diverse student populations

- Capture teachers' imaginations

- Are manageable in school classrooms

- Make use of materials that are inexpensive and widely available

- Facilitate integration, especially with mathematics and language arts

- Include methods for assessing student learning

Figure 4.2 Some characteristics of effective science materials. Adapted from Shuler, Sally G. (1993). Handout from 1993 Working Conference for Scientists and Engineers (March 22-26, 1993). Washington, DC: National Science Resources Center.

PART TWO:
A Session-by-Session Guide

Chapter 5
Preparing to participate

Chapter 6
Session I Activities and Topics

Chapter 7
Session II Activities and Topics

Chapter 8
Session III Activities and Topics

Chapter 9
Session IV Activities and Topics

Chapter 10
Session V Activities and Topics

Chapter 5

Preparing to participate

...most conventional educational researchers do not involve their subjects--the teachers--in the research itself. The teacher-subjects have little to say about the purposes, timing, methods, and tests involved.

... P. Blosser 1989

What are the distinctive features and components of the model?

This chapter provides a brief overview of the distinctive features and components of the process and how they are integrated. It also introduces the Session-by-Session schedule for the in-service experience.

The WyTRIAD has several distinctive features and components (Figure 5.1). Collectively, the **distinctive features** represent the integrated, structural and philosophical framework that makes this in-service model so effective. The **components** are the activities in which the partners engage. Each component is supported by both educational research and the experiences of teachers. The components are integrated and interconnected.

DISTINCTIVE FEATURES

Features that, collectively, characterize the WyTRIAD in-service model

- Partnership (teacher, building administrator, professional development facilitator)
- Long-term commitment and process
- On-site, in-context meetings with classroom-based experiences
- Based on the research on teaching and learning
- Constructivist approach to improving the teaching-and-learning process
- Constructivist approach to professional development
- Collaboration rather than just cooperation
- Teachers as researchers: inquirers of the teaching-and-learning process
- Modeling of all aspects, strategies, and components

COMPONENTS

Activities in which the partners engage during the process

- Interviewing students as a basis for instruction and curriculum development
- Experiencing and implementing a teaching for conceptual change instructional strategy
- Viewing modeled lessons
- Preparing and teaching sample lessons
- Collecting and evaluating data from classroom observations
- Maintaining reflective journals
- Peer coaching
- Collegial sharing sessions
- Reflecting on effectiveness of curriculum and instruction
- Reviewing curricular materials for appropriateness and effectiveness
- Making curricular decisions based on experiences with students
- Aligning curriculum, instruction, and assessment

Figure 5.1. Distinguishing features and components of the WyTRIAD in-service model.

How is WyTRIAD different from traditional in-services?

The WyTRIAD professional development model is a **synthesis of research-based ideas and strategies** that have been developed and tested by numerous investigators, along with some new ideas and approaches, combined in a way that is unique among professional development models (Figure 5.2). It differs from traditional in-service opportunities in several ways. Most significantly, it is an on-going process consisting of at least 5 or 6 sessions over a number of months. Some districts have chosen to repeat this process two or three times, and three districts have continued on with local teachers who are experienced in the model replacing the university researcher in the role of facilitator.

How is it different?

- It is a long-term process.
- It is a partnership.
- It is classroom-based.
- It applies a constructivist philosophy.
- Instruction begins at the learner's level.
- It emphasizes learning rather than teaching.
- It blends naturalistic research and implementation.
- It aligns curriculum, instruction, and assessment.
- Teachers become involved in research as a way to guide what they do.
- Teachers immediately implement ideas and determine what works for them and what does not.
- Teachers have time to reflect, ponder, share experiences, and question the appropriateness of assumptions made by textbooks and curriculum guides.

Figure 5.2. Some fundamental characteristics that distinguish the research-based WyTRIAD in-service model from typical staff development experiences.

It is constructed as a partnership, where cooperative relationships are developed. It is classroom-based, emphasizing learning rather than teaching, and begins where the student is in regard to conceptual development and prior knowledge. Conceptual change is the dominant theme of the WyTRIAD model. Participants learn how to promote conceptual change in the students they teach. On another level, however, the participants themselves undergo a process of conceptual change, changing their views of the teaching-and-learning process.

It integrates and builds upon the work of many educational researchers, and emphasizes teacher research as a basis for making curriculum decisions. It is a blend of naturalistic research and implementation that allows teachers a combination of advantages over other in-services. First, they can immediately implement ideas and determine what works and what does not. Second, they simultaneously have time to reflect, ponder, share experiences, and question appropriateness of previously unquestioned assumptions made by texts and curricula

When the Wyoming TRIAD program was begun, some of the reasons behind the present **lack of research-based teaching** were examined. Several issues came up, as elaborated in Chapter 1. It is useful to review these problems here.

• For the most part, educational researchers have not allowed the participation and involvement of the classroom teacher in the act of research and continue conducting research that can only be communicated to other researchers.

• As the subjects of research, the teachers have little to say about the research questions and how they are explored, even though the teacher's own classroom is the most natural, realistic setting for research.

• Upon close examination, we discover in many cases that educational research is narrow in scope, often flawed in methodology, and guilty of putting students or "subjects" into situations that do not realistically represent human behavior in natural settings.

• Teachers view most findings in educational research as impractical, difficult to interpret, and rarely possible to implement. Some researchers have interpreted this attitude as a lack of interest in research on the part of teachers.

• Another problem is in the way teachers are viewed. America is one of the few advanced countries where teachers have not achieved "professional" status. A common opinion in the general public is that research belongs to a select group and that classroom teachers are neither smart enough nor capable of doing research. Our society has accepted the belief that teachers should be viewed as "semi-skilled" workers who are kept so busy that they do not have the time to think and develop. We have approved the conditions under which teachers work. Many teach six periods a day and are required to supervise extracurricular activities in addition to preparing lessons and correcting papers after school.

• Traditionally, teachers have been expected to justify to researchers what they teach, how they teach, and how their students perform. However, we challenge this one-sided view. Researchers should have to justify to teachers and students their work, the importance of the research questions, their results, and the value of their findings.

These factors all have contributed to the failure of research to play a major role in professional growth and the improvement of professional practice. We believe that teachers need to study their work themselves, in their own way, and decide what changes they need to make in their classrooms.

This model was developed to help **overcome the barriers** to implementing research about the teaching-and-learning process. It helps to close the gap between research on teaching and learning and the implementation of that research in the classroom in two ways. First and foremost, it **creates confidence** about experimentation, educational decision making, and authority in the teachers, who after all, have the best environment for classroom research. Second, it works to bridge the communication chasm through collaboration. By collaborating with university faculty or other professional development resource persons, teachers have the opportunity to learn about and adapt effective educational ideas others have developed and to ask and pursue original questions about teaching and learning processes in their own classes. There are many possibilities, such as the few examples listed below.

• Teachers can investigate the effectiveness of a new program.

• They can study the appropriateness of certain concepts or objectives for a particular group of learners.

• They can study the effectiveness of such strategies as cooperative learning on students' achievement, social skills, and attitudes toward subject matter.

• They can investigate new assessment strategies.

This kind of opportunity builds confidence and professionalism as teachers find that their opinions are valued and that they are considered knowledgeable. Effective educational techniques from outside the teacher's classroom can be suggested by the facilitator. The teacher then can investigate the ideas and mold them to her/his own teaching style and the characteristics and needs of his/her own students. If the teacher has control over how and when particular strategies and materials are used, and sees results, that teacher will likely continue to use them.

What are the roles of the partners?

In the WyTRIAD partnership, the role of each of the three partners is equally crucial for success (Figure 5.3). Also, all partners are encouraged to maintain documentation of their observations and reflections about their experiences, thus being able to share and add to the cumulative understanding.

The role of the **teacher** is to initiate the concepts to be taught, use innovative teaching strategies, collect data, analyze and interpret data, and in light of what is learned, make judgments about the quality and the appropriateness of what should be taught and how. Additionally, the teacher is asked to step back and consider the meaningfulness of assessment and curriculum and work toward improving these based on her/his knowledge gained in the classroom.

The role of the **administrator** in this model cannot be understated. To maximize effective change in the classroom and curriculum, inclusion of the administrator is essential. As stated by Walton (1988), "No science education program can succeed without the support of the local school board and administration." Because administrators are the evaluators of teachers, policies, and methodology, they need to be knowledgeable of programs and curricular needs. Effective educational leadership, moving the group or the individual through the change process, is possible only to the degree that the principal understands how to facilitate it. Stated another way, the success or failure of any educational innovation in a school is directly related to the level of the principal's concern and actions. We believe that administrators hold the key to facilitating change in the classroom. Teachers require the leadership and understanding of the administrators in order to take risks using non-traditional teaching strategies and to grow professionally.

The **professional development facilitator** (university faculty or other resource person) provides expertise by helping teachers with interviewing, implementing the teaching for conceptual change strategy, peer coaching, collecting and interpreting data, and using the results in revising curriculum. The facilitator models the effective use of teaching strategies, lessons, and interviewing. It is also his/her responsibility to communicate research suggestions effectively, in accessible language. The teacher will have to have some knowledge about the research so she/he has the confidence to raise questions, challenge authority, and not accept new ideas on faith alone. This also means that the researcher must step back and allow some of his/her "territory" to become the teacher's, as the teacher becomes empowered

---

The classroom teacher

- Identifies the concepts to be taught
- Sets expectations for the students
- Determines student preconceptions in regard to the concepts
- Uses appropriate teaching strategies to create conceptual change in the learner
- Collects data
- Analyzes and interprets data and, in light of what's learned,
- Makes decisions about the quality and the appropriateness of what is taught and how it is taught
- Engages in peer coaching, sharing, and support

The building administrator

- Attends all sessions
- Enables teachers to carry out their role
- By being involved in the process and by understanding its rationale, provides the leadership, support, and flexibility to make it possible for teachers to pursue their mission

The professional development facilitator

- Provides expertise by helping teachers with
  - » interviewing students
  - » implementing research-based teaching strategies
  - » peer coaching
  - » collecting and interpreting data
  - » using the results to revise curriculum
- Uses questioning, explaining, modeling, and working interactively with teachers and administrators on site

Figure 5.3. Integrated roles and activities of the partners in the WyTRIAD in-service experience.

What is the impact of the different definition of roles on the teaching-and-learning process?

In the WyTRIAD, **all partners are learners**. Through becoming a researcher, the teacher, especially, expands his/her role as a learner. The role attains a dimension beyond what is typical of both traditional in-service education and traditional university course work. Also, some traditional aspect of each component of the teaching-and-learning process is **reversed or at least shared**.

- **Teacher**

Is the researcher, learning facilitator, and decision-maker.

Has a basis upon which to adapt or resist a curriculum that is imposed or dictated but that may be ineffective or inappropriate for her/his students.

- **Administrator**

Is the supporter, providing time, flexibility, substitute teachers, and encouragement, as well as being an advocate for the process and defending changes developed by the teachers when necessary.

- **Student**

Is the teacher, the one from whom everyone learns. In the WyTRIAD, the learner teaches us what is working, what is not working, and what changes we need to look at to make educational experiences meaningful.

- **Professional development facilitator**

Becomes not someone who pours knowledge on teachers but one who responds to the needs of the teachers and students.

Provides teachers with skills and strategies they need that they would not otherwise be able to get because of time or lack of expertise.

Models rather than directs, asks questions rather than gives answers, responds to needs rather than imposes directions.

- **Curriculum**

Changes, reflecting the dynamic nature of the process.

Rather than being totally accepted and implemented as developed by the district, it is subject to continued questioning, analysis, redefinition, and adaptation based on the classroom research.

Classroom research provides evidence for what does work effectively and appropriately in the curriculum.

Assessment moves from reliance on paper-and-pencil quizzes and tests for assigning grades to more varied ways of recognizing and promoting student growth in all domains.

What do WyTRIAD participants experience?

WyTRIAD integrates many features, some of which we will describe briefly here. Many of them are presented as "Topics in depth" in subsequent chapters. The core experiences for the participants are:

- **Modeling**

All strategies are modeled by the facilitator before teachers are asked to implement them, including interviewing, peer coaching, teaching using the conceptual change lesson, and effective questioning. The purpose of the modeling and the direct experience is to help <u>teachers</u> become comfortable with the principles and strategies to such an extent that they are able to fully implement them. It also is to help <u>administrators</u> to fully understand the rationale for the changes teachers are making and provide what is necessary to support the teachers in their efforts both during and beyond the in-service period. On-going, systemic change is thus fostered.

- **Sharing**

In regular group sharing sessions, participants report and share what they learn from student interviews, classroom observations, peer coaching, implementation of the new teaching strategies, and use of curriculum. These sessions develop mutual trust, confidence, and collegiality and strengthen understanding and respect between teachers and administrators.

- **Interviewing**

    Pre-instructional interviews are used to identify students' existing beliefs, understanding, misconceptions, and knowledge gaps as a basis for designing instructional experiences and making curriculum decisions. Post-instructional interviews are part of an assessment repertoire.

- **Teaching for conceptual change strategy**

    A six-step teaching for conceptual change model (CCM) is at the heart of the WyTRIAD. Teachers learn about the CCM and see it modeled by the professional development facilitator, both with the in-service participants as the learners, and with children in the participants' own classrooms. The teachers then design and teach lessons using this constructivist strategy.

- **Preparing and teaching sample lessons**

    Teachers prepare, teach, and evaluate lessons using the CCM in at least two subject areas. They also prepare, teach, and evaluate a lesson that integrates two or more subject areas.

- **Peer coaching and sharing**

    Peer coaching is implemented as an effective way for teachers to study their own mastery of specific strategies and to work collaboratively and constructively. Care is taken to design specific, non-judgmental, peer coaching experiences that are in character with the open, non-threatening, collegial atmosphere of sharing and working together. Mutual trust and respect are fostered.

- **Collecting data**

    Teachers collect data by interviewing their students, making observations of their students, and engaging in peer coaching and peer sharing. Recorded data are used as a basis for reflection and to document changes that occur in themselves and their students, classrooms, schools, and districts.

- **Reflective journals**

    Journals are maintained by all participants, in which they record both observations (data) and their analysis and reflections on what is happening. The cumulative and increasingly proficient journal entries provide a concrete basis upon which to build and improve professional growth and practice, document and justify changes, and share insights and experiences with colleagues and others.

- **Reviewing, evaluating, and making decisions about curriculum**

Textbooks, district curriculum, and other learning materials are evaluated based on what the teachers learn from interviewing and observing their students. Decisions are made about the developmental appropriateness of specific topics and materials, and these decisions are implemented.

- **Aligning curriculum, instruction, and assessment**

Emphasis is on the constructivist approach and the implementation of lessons that are modeled, developed, and evaluated; on the evaluation of curriculum; and on the collection of data through interviewing and observing students. These experiences are used to determine what is appropriate, accessible, interesting, and useful to learners at their various developmental levels, and then to bring the curriculum, instructional strategies, and assessment strategies into line with these findings.

Schedule and sequence

Figure 5.4 summarizes the scheduled sequence of activities and expectations that are set for each partner in this in-service experience. It is to be referred to throughout the experience, for implementing Chapters 6-10. The following paragraphs review, "in a nutshell," the partners' activities during the in-service experience.

At the beginning, **teachers** choose a science concept they plan to teach, develop a set of interview questions, then interview a sample of students to determine preconceptions about the concept. Teachers are often very surprised by the students' ideas, and express concern about the disparity between what they hear and what they thought they would hear. Next, the teachers examine textbooks and other teaching materials and compare the underlying assumptions and presentations of the concept to the children's ideas.

Teachers (preferably as teams) then prepare their lessons using the CCM. The conceptual change model involves small group collaboration and other classroom strategies. The formal cooperative learning strategy also works with the CCM, but the model does not require that teachers learn how to use it.

The lessons are then implemented. During each lesson, students and teacher work as a team. The students make predictions, explain them, and test them. The teacher facilitates, coaches, and helps students to experience conceptual change, as opposed to being just a "presenter" of information and an answer-giver. As they implement their ideas, teachers observe and collect data on changes they see in student attitudes, knowledge, skills, and interactions. They also collect data on changes they see in themselves, often with the help of a colleague through peer coaching.

Finally, using the insights they have gained from students and each other, the teachers evaluate their curricula. Their own data, analysis, and reflection are the basis for making decisions about the appropriateness of concepts, materials, and strategies for the children in their classrooms.

Because the **administrators** are with the teachers at all in-service sessions, they share in what the teachers have learned and understand why they are making their decisions about curriculum, instruction, and assessment. As key partners, they are aware of the rationale and of the new professional behaviors of the teachers and are able to support them in the changes they decide to make.

Throughout the experience, the **facilitator** informs, models, shares, listens, coaches, and advises. The experience concludes with planning for continued development and implementation of the ideas and skills that have emerged and been nurtured over the months of sessions and intervals.

Figure 5.4 (5 Parts). WyTRIAD In-Service Model: Session-by-Session Schedule

SESSION I: 1 day

Professional Development Facilitator	Classroom Teachers	Administrators
presents overview discusses what educators are learning from students examines students' cognitive frameworks and misconceptions presents & discusses interview process	participate in discussion	provide release time and coverage for classes for teachers participate in discussion
	initiate the concept(s) to be studied reflect on past experiences with teaching the concept(s) identify key points develop key questions for interviewing decide on expectations for students	listen to and encourage teachers
works with administrators for planning		participate in designing interviews and informed consent procedures prepare plan & list for teacher support between sessions

Between Sessions I & II: 2-4 weeks

Professional Development Facilitator	Classroom Teachers	Administrators
provides assistance	interview students review curriculum materials	listen to and encourage teachers provide teachers class coverage and time for conducting interviews provide recording equipment needed for interviewing
begins making journal entries: notes, observations, thoughts, feelings	begin making journal entries: notes, observations, thoughts, feelings	begin making journal entries: notes, observations, thoughts, feelings

Figure 5.4 (5 Parts). WyTRIAD In-Service Model: Session-by-Session Schedule

SESSION II: 2 days

Professional Development Facilitator

moderates and participates in sharing session

presents conceptual change strategy

teaches conceptual change lesson with teachers and administrators as students

models conceptual change lesson with students

facilitates discussion

provides expertise

works with administrators for planning

Classroom Teachers

share observations and reflections about students and selves

participate in discussion

participate in lesson and discussion

observe, record observations, and reflect

discuss observations

prepare own lessons using interview results and new teaching strategy

Administrators

provide release time and class coverage for teachers

participate in discussion

participate in lesson and discussion

observe, record observations, and reflect

discuss observations

participate and support teachers

prepare plan & list for teacher support between sessions

Between Sessions II & III: 3-5 weeks

provides assistance and expertise as needed

journal entries:
notes, observations, thoughts, feelings

use lessons in own classes

collect data on students' learning while teaching for conceptual change (journal)

keep track of changes in own philosophy and perceptions (journals)

support teachers
by providing planning time

make journal entries:
notes, observations, thoughts, feelings

Figure 5.4 (5 Parts). WyTRIAD In-Service Model: Session-by-Session Schedule

SESSION III: 2 days

Professional Development Facilitator	Classroom Teachers	Administrators
moderates and participates in sharing session	share observations and data about changes in students and selves	participate, support teachers, provide release time and class coverage
teaches another conceptual change lesson with teachers and administrators in student role	participate in activities	participate in activities
models conceptual change lesson with students in classroom	observe, record observations, and reflect	observe, record observations, and reflect
facilitates discussion	discuss observations	discuss observations
discusses and models classroom observation techniques	observe, discuss, and reflect	observe, discuss, and reflect
discusses and models peer coaching	observe, discuss, reflect, and plan for peer coaching	observe, discuss, and reflect
assists and supports teachers	prepare own conceptual change lessons based on their interviews of students	assist and support teachers
works with administrators for planning		prepare plan & list to support teachers between sessions as they plan, do CCM lessons, and do peer coaching

Between Sessions III & IV: 3-5 weeks

Professional Development Facilitator	Classroom Teachers	Administrators
provides assistance as needed	implement lesson plans	provide assistance and support
	observe students, especially changes	arrange coverage and time for peer coaching and planning
	engage in peer coaching with colleagues	provide recording equipment if requested
maintains journal	make journal entries: observations, reflections	make journal entries: observations, reflections

75

Figure 5.4 (5 Parts). WyTRIAD In-Service Model: Session-by-Session Schedule

SESSION IV: 1 day

Professional Development Facilitator

moderates and participates in sharing session
teaches conceptual change lesson in other subject areas
models conceptual change lesson that integrates disciplines with children in classroom
discuss observations
facilitates examination of curricula
assists as needed
works with administrators for planning

Classroom Teachers

share experiences and observations
participate in activities, reflect

observe, record observations, and reflect

discuss observations
examine curricula
develop own lessons/learning materials

Administrators

participate and support teachers, provide release time, class coverage, arrangements
participate in activities, reflect

observe, record observations, and reflect

discuss observations
examine curricula
assist in curriculum development
prepare plan & list for teacher support between sessions

Between Sessions IV & V: 3-4 weeks

provides assistance as needed

maintains journal

implement integrated lessons
collect data
implement peer coaching/sharing
review and reflect on own curriculum, instruction, and assessment (journal)
make journal entries: observations, reflections, especially in regard to C/I/A (curriculum, instruction, assessment)

support teachers to get materials

provide time and class coverage for teachers

make journal entries: observations, reflections

Figure 5.4 (5 Parts). WyTRIAD In-Service Model: Session-by-Session Schedule

SESSION V: 1 day

Professional Development Facilitator	Classroom Teachers	Administrators
moderates and participates in sharing session	share experiences and observations	participate and support teachers
	share reflections on evaluation of own curriculum, instruction, and assessment philosophy and practices	
discusses national trends	participate in discussion	participate in discussion
discusses strategies for curriculum revision	participate in discussion and planning	participate in discussion and planning

All partners discuss future plans, strategies, and grant opportunities to continue and expand the process.

After completion of all WyTRIAD Sessions

Teachers continue to develop and implement new lessons, revise curriculum, and engage in peer coaching, sharing, and other peer collaboration, with continuing support and assistance from administrators.

Seek opportunities to share experiences with other groups.

Continue to maintain reflective journals, including observations, ideas, thoughts, feelings.

All partners/participants stay in contact, especially in order to set up next cycle.

Professional development partner available for advice and assistance on concepts, strategies, presentations, expansion.

Chapter 6

Session I
Activities and Topics

Every man's world picture is and always remains a construct of his mind and cannot be proved to have any other existence.

... Erwin Schrodinger, 1958

The most important single factor influencing learning is what the learner already knows; ascertain this and teach him accordingly.

... David Ausubel, 1978

Learn about a pine tree from a pine tree, and about a bamboo plant from a bamboo plant.

... Matsuo Basho, 1644-1694

About the session

This chapter begins a series of five chapters, each dealing with a WyTRIAD session-and-interval pair. To be as helpful as possible to all of the partners, the chapters deal with both process and background information. In this chapter, Session I will be described in detail.

Often, Session I is a general in-service, following which the participants choose whether or not to become involved in the full WyTRIAD process. Alternatively, the participants may make commitments to become involved before Session I. In either case, Session I is the beginning of the process and thus sets the tone for the different approach used in the WyTRIAD model.

All three partners must take part in all sessions, beginning with the first session. Session I is scheduled to correspond with development time that the teachers have available, and with the availability of the professional development facilitator. It is equally important that the administrators of the participating teachers take part. The first WyTRIAD session defines the philosophical framework upon which the model is based, and <u>administrative support of teachers requires an understanding of this philosophical framework</u>.

The facilitator begins by introducing the other members of the WyTRIAD to the process. An overview of the WyTRIAD is provided first, along with some research-based statements about what we are learning from students. The facilitator then introduces the idea of student misconceptions, and provides some examples. The tone of this section of the meeting should be one of sharing stories about students, not one where knowledge is being handed down from a pedestal.

Teachers and administrators are then invited to suggest concepts that are relevant and timely to their teaching. From the group, a list of such concepts can be gathered quite quickly. In addition to identifying the topics, it is useful to come up with some of the difficulties related to teaching this concept. The group then specifically elaborates the expectations of students related to several of the concepts, and identifies key points regarding each concept.

Finally, these ideas are used to compose key questions to be used in structured interviews of a sample of students. The purpose of interviewing is to determine the students' entry conceptions and, quite possibly, misconceptions that relate to these concepts.

At this point, the facilitator models the interview process. The first of two or three students is placed in the classroom, where he or she can be interviewed by the facilitator while the teachers and administrators observe. (It can be quite interesting for the student to be asked questions by a stranger in front of a couple dozen teachers and administrators!) The interview typically takes ten or fifteen minutes and uses a few simple props. Following the interview, what was learned is the topic of discussion, and this can lead to more questions or commentary about the interviewer's choice of follow-up questions. A second or third student is interviewed, and then the discussion is reconvened to discuss the comparisons between the responses and their meanings. An alternative is to view videotapes of interviews with children, although videotaping requires certain precautions. (See Appendices regarding informed consent.)

The discussion of the interviews occurs on several levels. First, post-modeling questions regarding interviewing technique are answered. While interviewing well takes a great deal of practice, observing someone model the process is one of the fastest ways to learn. The interview discussion also touches on levels of understanding and misconceptions that were revealed, and how they are (or, quite often, are <u>not</u>) what the teachers expected. This is the second level on which this discussion can operate. Interviewing is a technique of getting much more in touch with what students believe and need to learn. Perhaps the most crucial point, however, is that the discussion of the interview modeling often leads to the revelation that some material that is being used is not appropriate for learners of a given level. Interviewing is shown to be an effective tool for determining what is and is not appropriate, and that teachers can be given credit, as professionals, to make decisions regarding appropriateness based on their own interviews and knowledge.

The final business of the first session is to discuss what teachers and administrators should be doing in the **interval before the second session**. This is, essentially, the "homework" for the in-service process, but, since the teachers have ownership and interest in the results, it does not carry the negative connotations of homework. During the interval:

- The **teachers** are asked to conduct individual interviews with several students on a concept of each teacher's choice. Again, the process is the same as that which was modeled: identify the concept; identify the difficulties, expectations, and key points related to the concept; then develop key questions that lead to the interview. Teachers are also asked to use the interview results to begin a review of the curriculum and its appropriateness for learners at their particular teaching level. Since interviewing will generally require teachers to spend time in ways not normally scheduled, they may need to make special arrangements. The arrangements will include finding appropriate times and places for the interviews. Covering classes for one another or some other strategy for making the time for quality interview practice will be needed.

- The **administrator's** role, therefore, is to support teachers by providing them with the time or the flexibility to make that time for themselves.

- The **facilitator** must provide time and be available for questions or assistance should that be needed between sessions. The facilitator also incorporates the teachers' concerns and perhaps the specific concepts suggested, into the planning for the next session.

Activities

Getting ready to interview

When preparing for the interview, keep in mind the following purposes for interviewing students.

• To identify learners' ideas and beliefs about different concepts

• To recognize the diversity of children's explanations

• To share examples of learners' explanations

• To identify and discuss contributing factors to learners' confusion or misconceptions

• To provide recommendations based on what is learned from the students in order to facilitate their construction of a valid concept

During the session, teachers, with the assistance of the other partners, prepare for interviewing in three specific ways.

• Identify key concepts.

• Prepare interview questions and think about potential follow-up questions.

• Determine methodology, including how to select students, setting, how to record responses.

NOTE TO FACILITATOR: Allow the teachers adequate time (perhaps an entire afternoon) to prepare and practice for the interviews, interacting with each other, the facilitator, the administrators, and other resources.

Topics in depth

What preconceptions, naive ideas, alternative frameworks, and misconceptions do learners bring to the classroom and why?

Students bring into the learning environment their own conceptions of the world. Their concepts often differ from accepted, scientific explanations such as those presented in textbooks and curricula. Several examples are presented below.

Example #1. Living and non-living

The concept of what it means for something to be "living" or "alive" is introduced in some curricula as early as kindergarten and is repeated almost every year. With each introduction, there is an increased expectation that the concept has been accepted. In spite of this attention, individual interviews of 30 fifth graders revealed that this is an unwarranted assumption (Stepans, 1985). Most of the students lacked a clear understanding of attributes of living versus non-living things.

Each student was asked if the following things were alive and why they thought so: sun, lightning, volcano, water, worm, leaf, tree, flower, candle, wind, bicycle. Students uniformly identified worm, leaf, tree, and flower as being living. The surprises came in their responses about the other items. Some sample comments:

Sun
16 students said the sun was alive. Some of the reasons they gave for their belief:
    it shines
    it moves
    it has energy which keeps us alive
    it keeps the solar system and the universe alive

Water
19 students said water was alive. Some of their reasons:
    it flows, moves
    it's made of living molecules
    it has insects, fish, bacteria, and plants living in it
    it gets hot, evaporates, goes to the sky, gets big, then gets
        down when wants to

83

<u>Lightning</u>
24 students said lightning was alive. Some of their reasons:
> it strikes
> it moves
> it makes noise
> it gives out light and can show you things
> it's partly fire and fire is alive because humans can't make it

<u>Candle</u>
14 students said the candle was alive. Some of their reasons:
> it has fire
> the flame is alive
> it burns
> when you light it you give life to it
> it's moving
> it's made of wax and wax was once a living thing

Example #2: Sink and float

Students at different grade levels were individually interviewed (Stepans, Beiswenger, and Dyche, 1986). The levels were primary (K-3), intermediate (4-6), junior high (7-8), and college. They were shown different objects and asked to predict if the objects would float or sink when placed on the surface in a container of water <u>and</u> to give reasons for their predictions. The objects were: large wooden cube, small wooden cube, small metal cube, looped wire, large metal cylinder, small metal cylinder, aluminum foil sheet, crumpled piece of aluminum foil, ball of clay, same ball of clay shaped into a pot or bowl, jar lid, and jar lid with holes.

The responses were grouped into three categories, based on how closely the student's explanation matched the scientific explanation, regardless of the vocabulary used. The criteria were:

> <u>Complete understanding (C)</u>
> Student was correct in the prediction and gave a completely correct explanation.
>
> <u>Partial understanding (P)</u>
> Student gave a partially correct explanation regardless of whether or not the prediction was correct.
>
> <u>No understanding (N)</u>
> Student gave an incorrect prediction and an incorrect explanation.

Surprisingly, the study revealed little difference in the understanding of most of the concepts among students at the different levels. One distinguishing characteristic among levels was the type of language students used. Elementary students used such terms as: weight, heavy, and light. Junior high students used such terms as: physical property, surface tension. College students used displacement, surface area, mass, and volume, but had little understanding of the concepts these terms represented.

Two examples of the results (expressed as percents) are shown below:

OBJECT	STUDENTS	COMPLETE	PARTIAL	NO UNDERST
clay bowl	Primary	0	89	11
	Intermediate	0	70	30
	Junior high	0	73	27
	College	6	79	15

A common prediction and explanation elementary students made about the clay bowl was that it will float because the water will hold it up. Older students commonly said it will float because we have made it lighter by changing the shape.

OBJECT	STUDENTS	COMPLETE	PARTIAL	NO UNDERST
jar lid w/holes	Primary	0	35	65
	Intermediate	0	15	85
	Junior high	0	7	93
	College	0	8	92

It is striking that the interviews revealed that the youngest students could give better predictions and explanations about the jar lid with holes in it than could the older students. Elementary students said the perforated jar lid will stay on top of the water based on their own, concrete experiences. They had played with similar things in the bathtub. Older students not only did not predict correctly, they explained that water will go through the holes and will make the lid heavy. They also used such terms as density and buoyancy in their explanations.

Example #3: Weather

Second and fifth graders were asked about specific weather phenomena--wind, clouds, thunder, lightning, rain, snow, and rainbows (Stepans and Kuehn, 1985). They were asked what they were, where they came from, and so on.

Their responses were placed into different categories based on Piaget's classification:

Feelings of participation:
Child feels personal action contributes to phenomenon.

Animism:
Child attributes life or consciousness to inanimate objects.

Artificialism:
Child thinks things happen for the good of humans and other living things.

Finalism:
Child thinks there is an explanation for everything, but that the explanations are not scientific. Rather, they are supernatural, with either a religious or non-religious component.

True causal:
Child gives an accurate explanation for the phenomenon.

The results of the study showed that some students used true causal explanations by fifth grade, but that both second- and fifth-graders exhibited predominantly non-religious finalism, such as:

Interviewer: *What is snow?*

7 year old: *There are white mountains where the white bears live and they would cut out snowflakes and then spread them all over.*

There were differences in the responses given by fifth graders and second graders on things like wind, clouds, rain, snow, and rainbow. For example, about 49% of fifth graders gave true causal responses for wind, compared to about 20% of second graders. On the concepts of lightning and thunder, however, there were no differences between grade levels; in fact, no students gave true causal explanations.

Example #4: *A Private Universe*

A videotape produced by Pyramid Film and Video, *A Private Universe* shows interviews with 23 Harvard graduates and faculty and a group of ninth graders. They were asked questions to determine their understanding of the causes of the phases of the moon and the seasons. While the Harvard graduates and faculty used sophisticated terminology in their explanations, 21 of the 23 showed a lack of knowledge and understanding of these topics. This was basically the same level of misunderstanding as the 9th graders. The videotape is readily available, and a similar videotape dealing with misconceptions in biology is presently in preparation.

What are we learning from learners about their preconceptions?

These examples and others test our assumptions about the teaching-and-learning process and about the science concepts children bring to class with them. We are learning from learners that:

- It is not valid to assume that learners have no prior knowledge relative to a concept.

- Young learners have their own theories and views of the world and these views may be quite different from those of one another and from adults.

- We cannot assume that when we present concepts in a certain manner the learners will immediately understand and accept them.

- Many adults, in spite of being exposed to many years of science instruction, tend to retain their naive ideas.

- Children's naive ideas may be similar to those held hundreds of years ago and documented in the history of science.

- Classroom instruction appears to be effective in bringing about conceptual changes with respect to some, but not all, science concepts.

What is the nature of preconceptions, naive ideas, alternative frameworks, and misconceptions and where do they come from?

All the ideas that students have, the ways they think, and the ways they view the world and things in it are carried into the classroom. Students are not clean slates, but bring their own understandings into the classroom. These pre-existing beliefs have certain characteristics.

- They are based on the children's own experiences.

- They are useful to and therefore valued by the children because they help explain children's experiences.

- They are very tenaciously held.

- They are often inconsistent with the scientific explanation.

As children form these untutored ideas, they are somewhat like scientists, in that they do collect evidence from their environments, and their minds attempt to understand and interpret what they experience. Their interpretations, however, reflect developmental factors.

- Their cognitive abilities are not yet mature.

- Their evidence is limited to their unaided senses.

- If they can't directly understand something, their brains will invent an explanation for it.

- Their ability to express their understanding is limited by their language development at any given point.

- They tend to be self- or human-centered and to project living characteristics and behaviors on inanimate objects and phenomena.

Some of the factors that contribute to students' confusion and misconceptions or that inhibit the alteration of their naive ideas are:

- Personal experience

- Language development

- The fact that students' ideas brought to the classroom are ignored

- Textbook presentations

- Teacher presentations

- Use of models

- Family and peers

- Media (including cartoons)

- Common sense

Recommendations to deal with students' ideas

A major goal of this project is to identify what **students** bring to the learning situation--their pre-instructional views of the world around them--so we can begin where the student is rather than where the curriculum or textbook is. We have identified interviewing as an important and effective vehicle to identify what students know and think about the topics that are to be taught. Interviewing also is effective in identifying students' confusion and misconceptions. One of the advantages of interviewing is that it reveals and allows us to clarify two separate things: **(1) what the learner knows or doesn't know, and (2) what the learner misunderstands**.

Interviewing also helps **teachers** to develop, both in instructional strategies and conceptual development about what they are teaching. They learn to formulate questions that are effective in eliciting children's explanations and develop their ability to ask meaningful follow-up questions to determine why students are saying what they say. Interviewing also helps teachers to identify gaps in their own knowledge and understanding of the concepts. This is often significant because studies have revealed that adults (including teachers) also harbor misunderstandings and confusion--about science concepts, in particular.

To begin to determine how to help learners overcome their naive ideas, we need to ask several questions, including:

- What do the students <u>really</u> know and think about the concept?

- How do we find this out?

- How do we incorporate what they bring into the class as we help them develop the concept?

- How do we use the ideas from textbooks and curriculum to help students develop a more accurate view?

- How do we help students to develop a more accurate version of the concept?

- How do we know if, in fact, we have helped students to develop a more acceptable view?

We know that students bring ideas with them and that it's important for us to use interviews and other ways to determine what they bring with them in order to know what to do about them. In some cases we may be dealing with pre-existing knowledge, but in other cases, with very young children, we may be dealing with things for which they simply do not already have a concept; e.g., numbers. The process by which they are initially introduced to a concept is critical, and we can't further a concept if the learner hasn't been introduced to it before. In helping students (1) to develop a conceptual change or (2) begin to process the targeted concepts, we should determine the appropriate starting point. Furthermore, by interviewing and talking to students we may find out that not only our assumptions about what students know <u>before</u> instruction on a topic may not be totally accurate, but also that we may be mistaken in our assumption of what they believe <u>after</u> instruction.

Another implication of research is that textbooks, other instructional materials, and the traditional way of teaching may not help students to overcome their naive ideas, and merit examination and analysis. Just as there should be good reasons and evidence upon which to base <u>changes</u> in practice, there should be good reasons and justifications for <u>continuing</u> to use specific materials, models, and strategies.

Teachers as researchers: inquirers of the teaching-and-learning process

Why should we research what happens in the classroom?

Suppose we have just begun to implement a <u>new</u> teaching strategy or activity in class. What information do we need to collect to see if it is working? How can we be sure we remember our observations accurately so we can analyze them as objectively as possible, based on the specifics of the observations rather than on generalizations or memory alone?

Perhaps more to the point, how do we justify what we are <u>presently</u> doing? How do we know if the present teaching strategies, assignments, activities, and ways of assessing are effectively serving our students? How do we know how well the teaching-and-learning process is working for <u>all</u> of the learners, with diverse learning styles and assorted similarities and differences? A Wyoming teacher shared the following reflection (Stepans and Saigo, 1993).

> *We wonder sometimes ... is this the time to teach this concept? Should it be some other time? I would like to see ... some determination at what level of mental maturity students have to be before you can effectively make changes in concepts. We can go over the same concept starting with first grade, and it's not taking there, and it's not taking in the second, and it's not taking in the third and fourth grades. Maybe we can put a little asterisk by this and say, "You might wait until the fifth or sixth grade before attempting to do*

anything with this concept because confusion continues to reign."

Most of us probably feel that we <u>do</u> adjust our behavior with learners in response to their reactions in the classroom; yet, in fact, most of us develop a pattern of teaching that is comfortable <u>to us</u> and then we stick with it. We review what the district curriculum requires of us, assemble our lessons, choose our materials and activities, collect attention-getting ideas at educational meetings (whether or not we are really successful at truly integrating them into our curriculum), set up our system of quizzes, exams, and grading, and then teach in fundamentally the same pattern year after year.

For various reasons, we settle upon a teaching style and repertoire of behaviors that suit us and that seem to suit a certain segment of our students--most likely those with particular learning styles that fit with our own teaching style. As the years go by, we tend to become increasingly confident of our personal systems. How many of us meet challenges to our established ways with statements such as "I've been teaching for 15 years and I know what I'm doing/what works/what students are like!"?

The organized body of knowledge about the teaching-and-learning process, however, grows constantly. It is our basic assumption, in the WyTRIAD, that this knowledge can and should be translated usefully into the classroom. A fundamental starting point is to become aware of both this knowledge and the methods by which it continues to reveal itself. One of these methods is organized observation of what actually happens to learners in a classroom setting. Making classroom observations, as we mean it here is, therefore, more than a casual, anecdotal phenomenon. We ask that teachers actively become researchers--that is, inquirers of the teaching-and-learning process.

Collecting data through classroom observation

As teachers teach kids, they collect information, or data. The information is collected by observing changes that occur in the classroom. What is observed? (1) What changes are seen in the way children learn, <u>and</u> (2) what changes do teachers see in themselves as they go through the teaching process? Ideally, during the WyTRIAD experience, teachers become increasingly self-aware about their data collection and classroom observation.

In the weeks between the in-service sessions, teachers listen to what the <u>children</u> in their classrooms say and the questions they ask, observe how children interact with one another and with the materials they are using, and interview them. They read children's assignments and tests. They also look for such things as attitudes, confusion, discouragement, apathy, enthusiasm, and success. Occasionally, they collect information from parents during parent-teacher conferences and other times.

In the WyTRIAD, teachers also are asked to become increasingly self-analytical. On their own, and with the assistance of their team members, they observe in <u>themselves</u> changes in the way they teach. They must, for example, actively analyze the kinds of questions they ask, the amount of talking they do, the amount of listening they do, and the way they incorporate students' ideas and allow them to share their ideas with each other and the entire class. Teachers also adjust their planning and instructional activities based on how the children are responding to the questions they ask.

Me? A researcher?

What do you visualize when you hear the phrase "teacher as researcher?" Does the term "research" evoke an image of the detached scientist performing complex scientific and mathematical rituals? The terms "research" and "researcher" may seem intimidating to those who don't do it on a regular basis, and bring memories of an educational statistics class, science laboratory, or reading a scientific article.

As it is presently used in science and mathematics education, the concept often refers to what is sometimes also called "action research." This term implies research that is driven by an immediate need and leads to some near-term practical application. So, we ask that you not be intimidated by the term "research" when we use it in this book. Perhaps different ways of expressing the concept would seem less formidable and would capture the essence more fully: teacher as explorer? teacher as investigator? teacher as inquirer, with research as inquiry? teacher as a constructivist of the teaching-and-learning process? teacher trying to make sense of what happens within students and then applying what they learn?

Now let's look at it from the other side. Can what teachers do during the WyTRIAD experience legitimately be called "research?" "Doing research" means **conducting an organized inquiry**. This operational definition applies to legitimate classroom-based research, research that has several characteristics.

- It has purpose.

- It has structure.

- It is more than making informal, scattered observations that are interpreted spontaneously.

- It takes into account published research and articles.

- It explicitly identifies and acknowledges variables and potential sources of error.

- It deliberately attempts to be objective and to minimize investigator bias, which is usually unintentional and unrecognized by the investigator.

- It applies appropriate analytical tools.

Teachers using the WyTRIAD model are required to conduct organized inquiry. They learn how to become researchers--inquirers--of the teaching and learning process.

- Teachers carefully design the questions they wish to explore, the methods they wish to use to conduct the inquiry, and the way they will collect and analyze data.

- They pay attention to the literature that illuminates our understanding of teaching and learning.

- At the same time, they realize that the purpose of their inquiry is to improve the effectiveness with which each of their students develops meaningful and appropriate concepts, skills, and attitudes.

- In this way, <u>their own research translates directly into better teaching and more meaningful learning</u>.

Interviewing

What do we mean by "interviewing?"

By "interviewing" we mean **conducting individual, structured, question-and-answer conversations** with a sample of students and recording the results of our interviews to establish a database for further reflection and action. Structured interviews are purposeful and planned. They are distinct from and more effective in probing students' ideas than individual or group brainstorming, casual questioning, and pencil-and-paper questioning.

In interviews we may be searching for several things. The scope of our search may be at the individual or group level, such as:

- A particular child's understanding of a topic or concept

- A better appreciation for children's science

- What kids think about things

- The diversity of beliefs about a concept held by the students in a classroom

When we analyze the responses from the interviews, we look for tendencies, as well as the views of individual students. Some of the inferences we can make about the group are:

- Individual patterns of understanding

- Group patterns of understanding

- Categories of responses that are related to levels of concept development

Why should we interview?

Whether we acknowledge it or not, our students already have beliefs and views about concepts that we are about to teach. In the past, we've used pre-tests or have simply assumed that we need to begin the instructional experience from where the textbook or curriculum is. We now are finding out that we need to begin where the student is. **If we do not acknowledge and incorporate students' ideas, meaningful learning does not occur.** That has been demonstrated repeatedly in formal research and in the informal findings of teachers. (Have you ever heard or said: "You should know this because you had it last year."?)

The best way to do this is to talk with students and have them share their beliefs and explain what they mean. Interviewing is a very effective way of finding out what students bring to the classroom. Because interviewing permits follow-up questions or "probes," we can elicit students' full personal explanations for natural phenomena, mathematical concepts, and relationships. Paper-and-pencil pre-tests cannot do this because they are not sufficiently open-ended and don't establish a friendly dialogue that permits probing for clarification, going both ways. Interviewing also can reveal student attitudes and other important variables.

Although conducting individual interviews takes more time than other means of eliciting information, the benefits more than outweigh the amount of time it takes. You might think of it as an off-setting "efficiency." The extra time it takes to interview a few students before you teach a concept means that time spent during the instructional experiences will be more effective for the teacher and for a larger proportion of the students in terms of meaningful learning.

Who do we interview?

As with other forms of data collection, it's important to attempt to get a representative **sample** of information. It isn't necessary to interview every student in a particular class. Usually, it is enough to interview 3 or 4 students. Ideally, the students should represent the diversity in the classroom, but it's also important that the teacher not "stack" the interview by deliberately selecting individuals. Some form of random or semi-random (stratified) selection of students should be used. For example, every 5th student on the class roster, or two each randomly chosen from a group of "higher-performing" and a group of "lower-performing" students would allow for some randomness and some diversity.

What can interviews do that other data collection methods can't do?

The main advantage of interviewing is that it provides information a teacher can not get otherwise. Interviews not only help to reveal what students bring to the classroom, but also help us to know if we are asking appropriate questions and if we are understanding students' responses. We can anticipate where and how pre-existing concepts--or their absence--may confuse or frustrate the learning process. Interviews can also help us to ascertain if we, ourselves, know the subject matter well enough to manage the follow-up questions that may arise.

Many teachers use an in-class questioning technique, but that has different purposes and outcomes than interviewing. Individual interviews are more effective in revealing what students really believe than large- or small-group questioning, which may tend to be dominated by a few students, intimidate more reticent students or those who are less fluent in expressing themselves, and be subject to social coercion.

How do we design and pose interview questions?

Interview questions must be carefully designed in advance and should be written down and rehearsed so the interviewer doesn't inject unintended meaning, personal bias, or confusion into the questions. If possible, have another member of your team or an experienced interviewer review them. Some sample interview questions are included in the Appendices.

Some points to keep in mind:

- When designing the interview, concentrate on a single topic or concept associated with the topic.

•	Summarize the key points you wish to explore.

•	Design key questions covering the points in your summary.

•	Make the questions as clear and straightforward as possible. Clarity is essential. Avoid asking compound questions.

•	Anticipate follow-up questions. You want to learn not only what the student thinks about a concept but also his/her explanation for that belief.

•	It is essential that interview questions elicit students' own explanations of their prior conceptions but don't put words into their mouths.

•	Questions must be open-ended and not provide other words to choose from. Written assessment tends to include vocabulary that prejudices students' explanations, whereas an interview with open-ended questions encourages free responses, those that are spontaneous and personal explanations rather than those limited to the use of a certain vocabulary.

•	Do not instruct, revise, or correct what the student says. This is NOT a "teachable moment." Anything the student says is to be accepted because it is what he or she believes and that is its value to you. You must resist the temptation to:

»	impose adult logic on the child to make sense of her/his explanations.

»	reword or rephrase student responses, or suggest a revised idea, as by saying "Then what you mean is...."

»	correct or clarify or give clues. Avoid prompting and leading questions.

»	react verbally ("that's good," "not quite," "why don't you try again," "well...") or physically in a way that would cause the child to know he/she has given a satisfactory or unsatisfactory response.

How do we analyze the information we get from interviews?

In reviewing the information, look for **patterns** of thinking, both of individual views and of the group. Responses of individuals may differ, but collectively they may converge on an idea; for example, where 3rd graders are in their understanding of weather. The interviews will give you an idea of the diversity of beliefs held by students in the class and their range of understanding of the concept.

Many WyTRIAD teachers find interviewing to be interesting and effective and make it part of their regular professional repertoire. The value compounds, in that it may take two or three years of interviews and looking at the literature and materials to get a fairly comprehensive grasp of what students know and think about a concept. The knowledge gained from year-to-year is useful not only in designing curriculum but also in justifying change.

How do we apply the information we get from interviewing students?

Information from interviews is useful at all stages of teaching a topic, from planning lessons through assessing learning.

- Interviewing helps us to identify an appropriate starting point. Rather than making a false assumption and looking to the textbook for leadership on the instructional level for a topic, we create our own assumptions based on the students' reality as reflected in the interviews.

- In our teaching of concepts we can refer to students' ideas, difficulties, and developmental levels that are revealed by interviews and adjust what we are doing. We can know to use alternative instruction and activities to help students construct a better understanding of the concepts.

- Information from interviews gives us a pre-instructional "snapshot" or "profile" of student ideas against which to compare assessments during and at the end of an experience or topic. It provides a baseline.

- We use interviews to judge the appropriateness of the curriculum rather than using the curriculum to judge the appropriateness of the students' understanding.

- The ability to explain and expand a response in an interview allows the student to reveal reasoning patterns and connections (or lack thereof). Instances in which the student links cause and effect, recognizes relationships among phenomena and between component parts of a system, and uses additional examples and metaphors are important data.

- Interviews can be used to evaluate the appropriateness of text materials.

- Interviews can be used as a basis for selecting and integrating various instructional strategies.

- Interviews help to increase awareness on the part of the students as well as the teacher, giving them a better sense of their own beliefs in regard to the concept. This is a direct benefit to the student in understanding his/her own learning.

- Interviews can help to measure whether or not conceptual change has occurred by comparing students' views before and after the instructional experience.

- Interviews can be a valuable assessment tool. Paper-and-pencil tests and other traditional forms of assessment primarily provide information that will tell us whether the student is "right" or "wrong" <u>regarding the specific question that is asked</u>. Interviews, on the other hand, give us an insight into what the learner is really thinking about a concept.

How do we prepare for interviewing?

- **Identify the topic**

Select a topic that's going to be taught in the near future; for example, magnets, weather, plants, simple circuits, multiplication, division, volume. Then clarify your thoughts about the topic. First, identify what concepts about the topic you are going to teach. Second, identify which concepts you think may not be appropriate for the level of the learners in your class. Third, explicitly identify any <u>underlying concepts</u> the students must understand to learn what you would teach and that you may be assuming they already know--your assumptions could be erroneous.

- **Summarize key points**

What is the scope and degree of complexity of the topic as you are going to present it at a specific grade level? For example, is the topic of electricity being taught to 2nd graders or 6th graders? For 2nd graders the key points may be: what is a complete circuit, what are its components, what happens when a circuit is completed. For 6th graders the key points may be: series and parallel circuits, conductors and insulators, short circuits, positive and negative charges, the relationships of current, voltage, resistance, insulators and conductors.

- **Design key questions**

Using the topics and key points, design key questions that will help you get at students' ideas about the topic. Think of 5-10 key questions to elicit natural responses.

For example, some key questions related to electricity and circuits that could be asked of 4th graders might be:

> "What can you tell me about electricity?"
> "What does it take to make electricity?"
> "What are sources of electricity?"
> "How is the electricity that we get in our houses the same as or different from the electricity that a flashlight uses?"
> "What are some of the words you associate with electricity?"
> "Have you seen a battery? What can you tell me about a battery? "
> "What does it take to get electricity from a battery?"
> "What does it mean to you when we say parallel circuit, or series circuit? How are they different?"
> "What does it take to have a series circuit? a parallel circuit?"

Questions can be based on the anticipated criteria. For instance, ask questions about relationships between components (the battery, the wire, and the bulb) such as: "What is the role of the wire when you hook it to the battery?" Or, if showing an apparatus, ask "Why is it important that the parts are hooked together the way they are here?" With your questions, you want to find out how the student links cause and effect in regard to a phenomenon or the student's understanding of the relationship between component parts of a system.

- **Effective follow-up questions**

We use follow-up questions based on what the students tell us to clarify what the student is saying without asking leading questions. For example, if the child says "Electricity in a battery is in the form of chemicals." We may ask "What can you tell me about those chemicals?" or "How do the chemicals make electricity?" A follow-up response can also be in the form of a statement, such as "Explain to me what you mean when you say...." or "Please draw a picture to show what you mean."

It takes a long time to become a good interviewer; that is, to learn how to design and ask questions, how to initiate interviews, to know what questions to ask and how to start questioning, how to ask follow-up questions, and how to respond to what students say, especially when their responses seem to be off the wall, or if students don't want to respond or say "I don't know." Be patient with yourself as you develop your interviewing skills. Remember that practice will improve them.

Some sample interviews are included in the Appendices.

Conducting the Interview

• **When do you do the interview?**

Do the interview a week or so before you begin any instruction about the topic so you can use the results of the interviews to help you design your lessons.

• **Why interview individuals instead of small groups?**

There is no substitute for individual interviews. In group interviews--even with only two students--students are influenced by other students' responses. Also, some students are eager to dominate the discussion and more reticent students' views will not be represented. To elicit the kind of information we are looking for, the interviews must be conducted with individual students <u>away from other adults and students</u>, and particularly not in the presence of parents.

• **How many interviews to do and whom to choose**

There are several ways to select students, but you want to assure a representative sample, as unbiased as possible by your own knowledge of the students or attitudes toward them. A sample of 3 or 4 students may be adequate. To make it random you can select every 5th student, or select one student from each level--high, medium, low--based on classroom performance. We are not trying to select students who can give the best answers. <u>This is a pre-instructional process so there is no concern for right and wrong answers</u>. We really want to know what the students think about the concept, what is their level of knowledge and understanding.

• **The setting**

One key to an effective interview is to be sure there is a comfortable, conducive atmosphere for the student. Before initiating the actual interview we need to make sure we develop trust with the student. Ensure that the student believes this is not a test, that it is not a matter of right and wrong answers, but that you are genuinely looking for what he/she really thinks. Put the student at ease by asking some casual types of questions at first; e.g., what kind of sports or other activities do they like, brothers/sisters, what kind of cartoons they watch, etc. It's essential that we convince the child that this is not part of a grade, and we won't be disappointed or be critical if she/he does not know any answers

Some characteristics of a good interview environment are:

- It should be non-threatening and risk-free.

- The interview should occur where other students or teachers can't readily eavesdrop.

- It should be quiet and relatively free of distractions.

- It also should be a place where the child can draw if that's required, or examine or manipulate physical materials if necessary; *e.g.*, which objects will float or sink.

- Keep it informal, friendly, and neutral. It is a conversation, not an interrogation.

- Establish the feeling of a dialogue in which you and the student are working together to explore a topic from the student's perspective.

- Be non-judgmental. You are not searching for correct and incorrect responses.

- Keep it non-instructional, as noted earlier.

- **Notes and audiotapes**

Since it is possible to miss quite a bit of information and subtle cues that could be helpful in understanding what the child is thinking about when taking notes, it is helpful to audiotape. Be sure that you understand the correct protocols that are necessary to protect the children you interview, as required by federal guidelines (next section).

Put the child at ease by asking permission, and explaining the reason for the tape recorder. You also must assure the child--and hold to your promise--not to use the recording in a way that the child could be identified or embarrassed. This assurance of privacy and confidentiality extends to your subsequent conversations with other teachers and students. Keep in mind that you do not have the right to play the tape publicly if the child's voice or name could be recognized without the child's and parent's signed permission (see Appendices).

Protecting privacy and the rights of human subjects of research

Is it necessary to get written consent for interviewing and taping? When it comes to research and personal data collection, everyone is protected by federal regulations. Minors, persons who are not competent to tend their own affairs, and prisoners have special protection as "vulnerable populations." For minors, both parental consent and assent of the minor are required for participation in research or data release.

If you are the one who is designing or administering some form of data collection, it is your responsibility to be sure that the rules are followed to protect the rights of anyone you collect information from and/or about, especially if their identities can be determined from the instruments and/or data. Audiotaping, videotaping, and questionnaires that require uniquely personal identifying information (name, ID number, or a specific combination of personal and demographic information) fall into this latter category.

Furthermore, schools--like medical facilities and other institutional settings--routinely keep confidential records about their clientele, including family information, grades and GPAs, test scores, individual evaluations, and health information. The data in these records is also protected and may not be shared without written consent, except when reported as group data in which the individual identity of any person cannot be determined.

These protections, however, do not paralyze your efforts to collect classroom data for use in instructional improvement and professional development. These protections are for you, as well as your students and colleagues. By following simple guidelines, you can respect and protect the rights of others and protect yourself from possible liability.

The federal regulations acknowledge that educational settings are special, because questioning and assessing are active and necessary parts of the teaching and learning process. The kinds of information-gathering normally done in classroom settings do not require written permission--they are considered to be an essential component of educational practice ("treatment"). The line is crossed when the information-gathering becomes part of a specific research project in which there is a research design that is not intimately tied to an instructional purpose and there is an intent to share more broadly what is learned. It also is crossed when audiotapes, videotapes, and photographs result.

What do you do, then, when you want to enlighten your own teaching practice by studying your students and sharing what you learn with colleagues? Figure 6.1 presents some simplified guidelines for gathering information from "human research subjects"--students, parents, colleagues, etc. As you read through them, please notice that they not only fulfill federal requirements, they also are a good vehicle for opening up and creating goodwill about the process. Greater detail is presented in the Appendices.

Protecting the rights of people who are studied

• Decide if it is important to know the identity of the persons you are interviewing or having complete a questionnaire. It is often important to know age, gender, and grade or educational level, but it is not usually necessary to have a name.

• If you are gathering longitudinal information to look for change (multiple queries over a period of time, pre- and post-treatment, etc.), you may want to create a confidential key sheet that will enable tracking of individual identities but that use numbers on the data sheets instead of names.

• Provide a brief explanation to the "subject" (student, parent, colleague, etc.). The explanation must include the following specific elements:

   » That you would like to ask some questions and collect some information.

   » The reason you are doing it.

   » Any risk or benefit that may be involved (*e.g.*, any impact on grade? any reward or privilege? any impact on instructional focus and activities?).

   » That the subject can ask questions about the session at any time.

   » That the subject can withdraw from the process at any time, without penalty.

• Obtain permission for the record. Sample consent statements are included in the Appendices.

• Be sensitive to and respect the feelings and comfort level of your subjects throughout the process.

• Keep your promises of confidentiality.

Figure 6.1. Some simplified guidelines for research involving human subjects.

Observation Notes and Reflective Journals

What is a reflective journal and what goes into it?

The WyTRIAD involves the partners in several strategies and simultaneous growth in various directions. Because it is intensive, it may be easy for participants to overlook or forget important information about their experiences and their students' experiences. So, by keeping journals partners write down evidence (data) and their subsequent reflections and analysis of this information.

The journal is a **professional tool**, not a term paper or project. It should always be regarded as a "working draft," and it is not necessary to compose outside the journal, then rewrite into the journal. It includes useful notes, observations, and reflections based on observations of students, those special moments or things that happen to particular students and to the person writing the journal as a result of these experiences, their reflections on sharing sessions, their beliefs, their attitudes toward peer coaching, interviewing and what they have learned from talking with their students, impressions, ideas, and connections. **Teachers keep reflective journals to document and analyze the changes that are occurring in themselves and their students.**

Data in the journal include such observations as: What were the children doing during today's class time? Were they expressing ideas? listening to the ideas of others? interacting and sharing ideas? asking questions? talking? laughing? Was there a lot of activity and emotion, from which we could infer they were enjoying the activity? Did they seem to be intensely absorbed in the activity? Did the specific activity engage any students in a different way than usual? Were there frustrations?

Reflections in the journal include inferences or opinions after thinking (reflecting) about what was observed or experienced. Are the students enjoying the activity? Do my observations reveal that many of them are confused? understanding? Am I able to tell if learning is occurring? How does this group of children differ? How am I as a teacher different when I'm using the CCM? What are my comforts and discomforts, thoughts and feelings, successes and "do-betters?"

The **physical characteristics** of the type of book to use for a journal vary among teachers. Often, a spiral notebook is used. For permanence, durability, and ease of shelving, a regular, bound composition or research journal is very good. Some teachers use a stenographer's pad. A common, required feature of all of these types is that they are bound, not loose-leaf. Some people find they are more fluent using a computer to develop their reflections.

The **format** for recording information should be comfortable and convenient for the individual, and lend itself to easy reference for subsequent development of reflections. One format that has been used and is very easy to follow divides the page into two columns--a steno pad format (see Appendices). On a given day, the teacher makes notes and records observations in the left column. As she/he thinks about the observations at any time, reflections can be written in the right column, near the information upon which the reflections are based.

How is a reflective journal different from a diary?

There are distinct differences between a reflective journal and a diary. A journal includes evidence and analysis, and it may also include reactions and feelings. A diary is often a simple chronicle that includes reactions and feelings. For example, a diary entry might read "Today we did [...] and the kids really liked it." That is a subjective generalization. It does not contain any data to support its inference.

A reflective journal is not just a record of what happened and what was observed, however. It is a place to verbalize new things to be tried, in which teachers set expectations and make predictions about what they think will happen when they teach the lesson or work on a new strategy. They also write down what happens--both expected and unexpected--then analyze, summarize, and reflect, and even extend to further questions and ideas. In this way, the thought processes captured in the journal mirror some of the stages of the conceptual change teaching model, but in a personal, metacognitive way. It puts the participants into an active role of being learners about their own professional activities.

Data should be recorded directly into the journal, then thought about later. This is an ideal way of using the journal and is a parallel to the way a traditional laboratory notebook is kept--a combination of data, sketches, ideas, hypotheses, inferences, and thoughts for future work. It also can include what may be thought of as "teaching notes," for future reference. You might record, for example, what worked well and less well, what additional materials or activities would you add, what would you do differently with the activity next time, whether or not the students seemed to grasp what was expected. Many teachers keep such notes in their lesson plans, although some have commented that sometimes the meaning of the note is obscure by the time the next year rolls around.

The example presented in Figure 6.2 is a hypothetical reflective journal entry that includes several of the elements mentioned above. Because it is hypothetical, the language is probably less spontaneous than a real journal reflection would be, but it does demonstrate a combination of exploring new ideas, gathering information about their implementation, and reflecting about what is observed. Many entries will be much briefer, such as the hypothetical example in Figure 6.3. Although it is brief, this entry also contains both data and inference from the data.

Why is it important to keep a journal?

Keeping a reflective journal is a significant component of the constructivist philosophy in the WyTRIAD. It is a tool to help participants begin to develop a constructivist view of the teaching-and-learning process. It is a record of how participants actively and deliberately challenge their own pre-existing beliefs and construct their own new knowledge about teaching and learning.

The journal contains evidence upon which the participants make informed curricular decisions. As such, the journal can be used as a basis for further growth, curriculum revision, selection of materials, presentations, and publications. It is an informed, useful document that is valuable for sharing real information and rationale for change, as with other teachers, parents, and school boards. If things are not written down, the details are easily forgotten or misremembered. As Benjamin Franklin is credited with saying, "The palest ink is better than the fondest memory."

Use of journals is popular with many in-service programs, but it is also one of the most troublesome things for teachers. Often participants are not clear about what the journals are, how they are to be maintained, how they might be useful, how much time they might take, and so on. They worry about "being graded" on the journal, and fuss that entries must be drafted and edited before they are inscribed in the journal--a lot of effort. Also, teachers may not be comfortable having others read their journals, such as an instructor or professional development facilitator. As a result of all of these perceptions, journals are often the first component to be dropped when outside pressure is gone.

In the WyTRIAD we attempt to overcome these barriers and emphasize that the journal is a professional tool to be used "constructivist-ly" by the individual, that it is not a term project, that it combines spontaneity, reflection, and utility, and that it is also not just a diary for chronicling events and/or expressing personal feelings.

Class: *Life Science*   Date: *October 12, 1995*

When I got ready to teach about insects in the past, I set these expectations of my students--that they had to learn a list of terms and facts, do experiments with mealworms, collect and classify 25 insects. As a result of my new thinking, however, I realized that these expectations are unrealistic for my students and, more importantly, may not give them the understanding about and appreciation of insects I'd like them to have.

In the past, I usually organized all the experiences around an initial lecture about insect anatomy, diversity, and kinds of life cycles, defining the vocabulary first, then having the students do worksheets and lab work. This time I interviewed some students before beginning the topic. I found holes in their knowledge, but also some things they seemed to know. I also got ideas about their attitudes toward studying insects--bo-ring. As a result, I expanded my plans to include more environmental aspects and human-insect interactions, deleted the dissection of a preserved grasshopper, and focused on involving the students in field studies.

I also got an idea for doing the anatomy part. Many of the alien creatures in cartoons and movies are insect-like in some respects; so, to start the unit I showed pictures and video clips of popular fictional creatures and various insects and asked the students to identify what it was that made the aliens insect-like, as a lead-in to defining the construction of the insect body. Their observations led right into an analysis of the relationship of structure to function.

We also brainstormed about kinds of insects, where they live, and the roles they fill in natural food chains, then set up explorations to find insects (including field trips and figuring out ways to capture insects in different kinds of habitats and at night). We developed the terms and concepts as needed to explain the students' findings and applied them to other concepts, such as reproduction and life cycles, defense mechanisms, ways they compete with humans for food (and use humans as food!), how they are beneficial, and how they also can be vectors of disease.

We ended up taking longer on insects than I had intended, but the students were so engaged in it that I used it as a way to introduce and reinforce other basic biological concepts and principles, in the context of the unit. We also got into a little biophysics, exploring the different kinds of sounds insects make and how their appendages work.

Figure 6.2. A hypothetical journal entry that reflects changes in philosophy and instructional strategy.

Class: Life Science   Date: April 6, 1995

The noise level varied during the class period. In the beginning the students talked in moderate voices and moved about as they began to make observations and compare the spots of bacterial colonies and fungi that were developing on their Petri plates. It even got quieter as they intently measured and recorded observations. While they were analyzing the data collected over several days, however, the noise level increased. I think the students were excited about the results of their investigations.

Figure 6.3. A hypothetical journal entry that, although brief, contains elements of both data collection and inference from the data.

Between-the-session activities

Interaction of the partners

During the interval between Session I and Session II, **teachers** are to implement some of the activities, especially the interviewing, and be prepared to share their experiences with the group during Session II. The critical supportive role of the **administrator** is listed on the Session I page of the Session-by-Session schedule (Figure 5.4). In addition, the administrator should record her/his observations in a journal to help participation during session sharing, as well as to support reflections about changes in the teachers, school, and self. The **facilitator** should provide the teachers and administrators information about office hours and availability for visitations, try to make informal observations of teachers and administrators between sessions and keep notes of telephone, email, and other contacts in her/his journal.

Specific instructions for teachers

- Interview students (Figures 6.4 and 6.5)
- Compare students conceptions with curriculum (Figure 6.4)
- Make daily journal entries

Reports on interviews may be prepared and shared by individual teachers or by teams of 2-3 collaborating teachers.

Provide the following information:

1. Topic chosen
2. Summary of the topic
3. Expectations you set for the students (include skills, content, and attitudes)
4. Key questions you used
5. Sample of students' responses and your follow-up questions
6. Synthesis of the interviews (what did they reveal?)
7. Examples of interesting and unexpected responses
8. Comparison of children's views and the assumptions made by your textbook or other instructional materials
9. Implications of all of the above to curriculum, instruction, and assessment
10. Overall reactions to interviews and their uses
11. Additional comments

Figure 6.4. Specific instructions for recording and synthesizing interviews to be completed before the second WyTRIAD session.

Becoming a good interviewer takes practice

1. Begin the interview with an explanation of purpose.
2. Record the interview with a tape recorder.
3. Be a good listener.
4. Encourage a response rather than accept "I don't know."
5. Use spontaneous probing questions when necessary.
6. Remember: this is <u>not</u> an instructional situation.

Figure 6.5 Hints for beginning interviewers.

Specific notes for the facilitator

1. Be sure to have large sheets of paper (such as butcher paper or easel paper), markers, and masking tape at <u>all</u> sessions.

2. Also, the facilitator needs to be prepared to facilitate the sharing process that begins Sessions II through V through the anticipated and well-documented ups and downs of group dynamics and change.

3. Prior to Session II the facilitator prepares a conceptual change lesson to be modeled.

4. Regarding time requirements for developing lessons: During Sessions II, III, and maybe IV, give teachers free time all afternoon to work on developing lessons. They will need to use curriculum materials and other resources, consult each other and the in-service facilitator, and try out materials. See Session-by-Session schedule (Figure 5.4).

Specific notes for the administrator

1. Work with the teachers and facilitator in advance of Session II to arrange <u>space</u> at school and to <u>adjust class scheduling and coverage</u> as needed to accommodate the activities, some of which take place with students in the regular classrooms.

2.   Regarding time and schedule to plan for modeling: See Fig. 7.1, "Ideal schedule for modeling day." The most effective way to manage this is to <u>release participating teachers for the entire day</u> so they can fully participate in the experience. This day is as much a part of their assignment as if they were in class away from the school for the entire day--consistent with the idea of treating teachers as professionals. It is important that they be able to devote full attention to this cohesive experience.

Chapter 7

Session II Activities and Topics

Peer interaction is more effective than teacher lectures in conveying scientific knowledge because peers' explanations are simpler than adults' and as a consequence are better understood by the learner. ...the distance between two student's understanding is far less than the distance between a student's understanding and a teacher's, hence communication of ideas is facilitated. Often a peer is quicker to identify a point of confusion than a teacher.

... Audrey Champagne and Diane Bunce, 1991

About the session

The beginning of the second session and each following session should be set aside for **sharing**. This sharing includes not only observations about students and what was learned from them, but also what the teachers learned about themselves. Begin the sharing in small groups, where teachers can spend some time reviewing their written observations and sharing stories about what happened in the time that passed since the last session. They also should express concerns and problems. Fifteen to twenty minutes is a useful amount of time to spend in the small groups.

When the discussions and sharing have reached a natural conclusion, some of the new topics for the second session can begin. The main goal of the second session is to have the participants experience the conceptual change strategy and develop an understanding of its rationale, pattern, and effectiveness. Fig. 7.1 shows how the session could ideally be structured.

Next, teachers and administrators experience a lesson designed in the format of the Conceptual Change Model (CCM, Chapter 3). The conceptual change strategy is effectively introduced through modeling. The facilitator models this strategy using a topic that is typically in the curriculum and using materials not unlike those the teachers might use in their classes. Teachers and administrators play the role of students in the initial modeling session and learn how each step of the strategy leads to the next. As students, the teachers and administrators are expected to get involved in the activity by openly sharing personal predictions and ideas.

This activity, when nicely paced, can lead to an effective discussion about how students learn. What teachers learned from interviews can also be incorporated into the discussion. Since this is the first opportunity to integrate some of the main components of the WyTRIAD, the interrelationship can begin to be constructed.

The next day, the same CCM lesson is modeled twice, with students at two different grade levels, preferably in the teachers' own classrooms, with teachers and administrators observing. After the lessons have been modeled, there is discussion (no students present) about what the participants observed and the educational implications of their observations.

At this point in Session II, time is set aside for **teachers** to work in groups to decide how best to apply this new teaching strategy to their own teaching. Based on their interviews, their experiences and observations of the modeled lessons, and the ideas and experiences that have been shared by their colleagues, the teachers develop their own CCM lessons. They also use this time to get assistance from the other partners.

During this work time, the **facilitator** provides expertise as needed, but generally the teachers are free to work the six steps of the strategy into their teaching of a concept as they see fit. This freedom is provided so the strategy is most likely to be utilized effectively by the teacher, which only occurs if the teacher has multiple opportunities to develop ownership of the strategy and its underlying concepts.

The **administrators** have been observing and participating in the session to this point, and here it is important that they express some support for experimentation. While taking a whole new approach to a concept might constitute a risk for some teachers, the support of the administrator can reduce that feeling of risk quite substantially. The focus of the change in teaching strategy can then be on improving the instruction of the students, where it belongs.

The final part of Session II is to clarify the assignment. It involves talking about what the teachers should try out during the weeks between Session II and Session III.

Ideal format for days in which
the conceptual change model (CCM) is modeled in classrooms

THE DAY BEFORE

The lesson demonstrating the CCM is
experienced by teachers and administrators.

THE DAY OF THE EVENT

Teachers are freed from all classroom responsibilities for the day
to participate. Administrators are scheduled to participate as well.

Early morning

1.   Teachers, administrators, and facilitator meet as a group for pre-instructional phase of the modeling.
2.   Lesson is modeled by the facilitator in a classroom, with students.
3.   Post-instructional discussion, analysis, and reflections.

After lunch

4.   Teachers, administrators, and facilitator meet as a group for pre-instructional phase of modeling in a different grade-level class. The only change here is discussion of any changes to be made in implementation of the lesson based on what the teachers and administrators observed during the morning lesson.
5.   Lesson is modeled again by the facilitator, but at a different grade level.
6.   Post-instructional discussion, analysis, and reflection. In addition to analyzing the second lesson, generalizations are made based on observation of both classes.

After school or in the evening

7.   WyTRIAD session activities continue as per schedule for specific Session.

Figure 7.1. Ideal format for instructional modeling days. The administrator is responsible for making all space, student, and personnel arrangements in advance, coordinating with the facilitator.

Activities

Share interview results

How to share

Teachers get into groups of 4 or 5 individuals who teach the same or close grade levels to share in small groups the results of their interviews in regard to the specific questions listed below. The administrators participate in the small group sharing. After successes and concerns are discussed by each small group, a spokesperson from each group can share them with the larger group, where discussion can be helpful. Every teacher should share something for discussion and actively take part in the discussion.

It is crucial that the facilitator explicitly remind participants to <u>keep open minds</u> about what they hear during the sharing at each session. All colleagues have valid experiences and valuable ideas to share. Since this is a process involving change and many new ideas, there may well be personal feelings of discomfort and even some resistance.

What to share

Teachers bring their observation notes and journal entries with them to use during the sharing session. They should be prepared to share the following:

- What concept did you choose?

- What were the key points you wanted to cover?

- What were some of the key questions you selected?

- What are some sample responses from students?

- What are sample follow-up questions you asked?

- What were examples of interesting or unexpected things your heard from the students?

- What was the diversity in responses you encountered?

- What is your synthesis of what you learned from the students?

- How did students' responses compare with the assumptions behind how the concepts are presented in your textbook?

Summarizing the sharing

In the small groups, record some of the key points or ideas of particular interest on a large sheet of paper for sharing with the larger group later. After each teacher has had an opportunity to share the results of her/his own interviewing, discuss in the small groups what are educational implications of what was learned from interviewing students, and your impression of the value of the interview.

After this discussion, a spokesperson from each small group presents a summary of their sharing to the entire group.

Introduction to the Teaching for Conceptual Change Model (CCM)

CCM modeled by facilitator

At this point, the teachers have identified their students' preconceptions about the selected concepts. The next effort is to help teachers create a conceptual change in their students in respect to the concepts. The starting point for this effort is for the facilitator to model a conceptual change lesson for the participants, with them in the role of students. The strategy we use is the six-step conceptual change model for teaching and learning (Stepans, 1992, 1993, 1994) presented in Chapter 3 and Figure 3.1.

A lesson we have used effectively for this modeling deals with the behavior of pendulums. The lesson is reprinted in the Appendices, with the permission of the publisher. You will note that this lesson consists of a set of activities (experiences, or situations) that collectively include all six stages of the CCM, although each activity within the topic may not require all six.

After the teachers and administrators have experienced the lesson, they analyze their responses to it and discuss how they think students will respond. Then, the next day, they see this same lesson modeled by the facilitator with children in two different classrooms at two different grade levels. Generally, this is done with a participant's own class. The facilitator can usually adapt the selected concept to whatever age of student is available. It is beneficial to model the instruction of this concept to two different age groups, one after the other, if such a situation can be arranged. Again, following the observation of these two modeling sessions, a useful discussion can be continued about what was seen and learned.

As the lesson is being conducted, teachers and administrators are asked to observe and take notes about it. They should particularly observe:

- How do the students react to the model compared to how the adults reacted when it was modeled for them?

- What questions are the students asking?

- What common misconceptions are revealed?

- How are the students interacting with each other?

- Do any changes seem to be coming about as a result of the teaching strategy?

Reflection and discussion about the lessons

After a short break, the group reconvenes in another room to discuss the modeling. The facilitator needs to be sure all participants contribute to the discussion so the group will have the perspectives of all of the teachers and administrators.

One of the burning issues may be a discussion of common misconceptions about the behavior of pendulums or other concepts that were modeled, especially since those ideas are held by many adults and are actively fostered by many available curriculum materials.

Based on the observations of the participants, the discussion will focus on the students' responses to the lesson and the similarities and differences from when the teachers and administrators went through the activities. What were the adults' reactions to the teaching strategy compared to students, in general, and in the different grade levels? What are strengths and limitation of the teaching strategy as they observed it? Also, what are the implications of this discussion to the teaching-and-learning process and to the school; for example, to restructuring the schedule, learning environment, and curriculum? It is critical that administrators be a part of this synthesis of observations and discussion.

**Begin to develop learning activities
based on interviews and observations**

At this point, teachers prepare to develop their own lessons. To put it in a more constructivist way, they start developing **appropriate experiences** for their students based on: the results of their interviews with students, discussion with colleagues, their own experiences and responses to the CCM lesson, and the modeling of the strategy with students. In preparation for developing lessons, teachers have already done several things:

- They have identified topics to teach.

- They have identified key concepts to be developed within those topics.

- They have interviewed their students to find out their views and misconceptions.

- They have reviewed how the concept is covered in existing curriculum and textbooks.

- They have taken time to think about and recognize the level of their own preparation on the concepts.

- They have shared their thoughts on the match or mismatch between what students say and the assumptions and expectations that are set for them by the curriculum and textbook.

Now they set their own **expectations** for students related to the concepts. These expectations should cover not only what students should know and be able to apply about the concepts, but also skills and attitudes the students should develop. Based on the expectations, teachers use the results of the teaching strategy they have experienced and seen modeled and decide on appropriate **experiences** in and out of class for the students. In addition, teachers develop appropriate **assessment** strategies (interviews, observations, tests, performance, etc.) to determine whether or not the expectations of students are met. The result, from interviewing to development of curriculum, instruction, and assessment, is a coordinated stream of events:

research —> expectations —> experiences —> assessments

Some planning and format ideas for the lessons are provided in the Appendices, along with sample lessons in different content areas and assessment ideas. The pendulum lesson presented is well-tested and may be a helpful model. It was developed through the research-based process, including student interviews. As with the rest of WyTRIAD, we hope to inspire and encourage a fresh approach, freeing teachers from whatever conceptual block or preconceptions there might be about lesson planning and "formula" or "prescription-type" lesson plan formats. Instead of going by restrictive, closely formatted lesson plan outlines, we encourage teachers to rely on themselves and their colleagues as professionals to come up with a newer, more flexible, and more fundamental kind of plan that is based on their own classroom research. Through practice, teachers usually discover how to use the new teaching strategy in a way that is comfortable for them.

Between Session II and Session III, teachers implement the lessons they design in their classes. They will collect data on the effectiveness and appropriateness of the materials they have developed. Collegial collaboration is recommended.

Topics in depth

Sharing

How important is the sharing component?

Many new things happen in this WyTRIAD project. Teachers are introduced to new concepts and strategies, doing research, keeping journals, etc. The new instructional strategies work differently with different students and different concepts. Sharing is one of the most important components of this experience, providing all partners an opportunity to learn from one another and develop common understandings.

Sharing that occurs during the sessions is an opportunity for teachers to bring to the group what they have done in their classrooms--successes, things that haven't worked, frustrations, problems, excitement, enthusiasm--so they can learn from one another. They share a wealth of information and ideas with one another and give and get practical, useful feedback. Differences in perspectives across grade levels provide longitudinal insights into the appropriateness of concepts and activities at various grade levels. Collegial relationships develop based on opportunities to assist and reinforce one another, sharing similar frustrations and successes. As important as the sharing process is, we must keep in mind that we are encouraging teachers to participate openly in these discussions, not forcing them to "make reports," which diminishes or reduces the sharing to a dutiful activity rather than a creative, spontaneous activity.

Through sharing, facilitators who are staff development persons or university researchers find out what is really happening in the classroom. They have opportunities to link the research literature and national recommendations with which they may be familiar to actual experiences of teachers. Also, the experience is a source for new research ideas, generation of questions, and development of research collaborations with teachers.

As all the partners become confident in the activities and philosophy, and certainly after they have been through one full in-service, they are encouraged to extend their sharing about what they are learning from students and their own experiences to others <u>outside</u> the WyTRIAD group. Other administrators and other teachers in the building find out what the teachers are doing and have opportunities to see how classrooms (and students) can be transformed. Parents also are part of the information network, through their children, projects, parent-teacher conferences, and PTA/PTO presentations and newsletters. Local news media even may become involved; for example, doing a story on the fun students are having in science class, or a project they may be involved in.

Who should be involved in the sharing process?

It seems obvious that teachers should share their experiences, but none of the partners should be excluded from the process. Filtering the experience through all the different perspectives helps to get the most out of it and provides the best advice. It requires effort to create opportunities to share--especially for the involvement of all three partners, not just the teachers. Sharing as a form of outreach can be very useful, bringing others, such as prospective teachers and parents, into the sharing process.

What does collegial sharing accomplish?

If we want teachers to change, take risks, and try new things, it's important for administrators to understand what the teachers are doing and to actively provide a support system. This support extends beyond the in-service itself. For instance, if other people from the district, parents, or school board members are questioning the changes, the administrator who has been a part of the process understands the rationale, philosophy, and activities and can fully explain it to support the teachers. If administrators are not involved in the sharing, they don't have a solid idea of what is going on and are less able to support and perhaps even to defend what the teachers are doing.

Because they are part of the experience, administrators can be more sensitive to the teachers' need for time to collaborate, be creative, and implement changes to improve the experiences and success of their students. If teachers elect to use alternative assessments and new teaching strategies, work with other colleagues, interview, be involved with peer coaching, or even need to travel to work professionally with other schools, etc., the administrator is the key person to enable these endeavors. Participation in the sharing of observations and the applications of these classroom observations, causes administrators to be familiar with the evidence and understand the rationale for what is being done.

Another way sharing is useful to teachers is that as they plan a new unit, they can get ideas about what other teachers have done to be successful and avoid problems or ineffective materials and activities. Sharing encourages feelings of professionalism, that they belong to a community of colleagues who are searching, growing, and trying new things.

Through contact with university and other professional development persons, teachers are in touch with national movements, opportunities, materials, and relevant research. Extending further, through sharing ideas teachers get new ideas to bring to and from school and district curricula. All of this communication helps teachers to be aware of what's going on at many different levels. For example, WyTRIAD teachers who have become involved in planning curricula find that they have a larger overview of curriculum, that their knowledgeable views are respected in the process, and that they have a basis for deciding that specific topics and activities are included in the curriculum because they are appropriate and representative and not just because they are someone's favorite lesson (Stepans and Saigo, 1993).

What kinds of things are important to share?

With the schedule teachers have, they seldom have the opportunity to share with colleagues some of the exciting things they are doing in their classes. Nor do they have opportunities to seek help with areas of concern in a safe, mutually supportive atmosphere. Faculty meetings normally deal with the business and administrative aspects of school. WyTRIAD provides the teachers and administrators the opportunity to share, expose their frustrations, needs and possibly inadequacies in a safe environment, with their colleagues.

After the opening session, each subsequent session begins with sharing. Teachers usually come to the sessions excited about telling their colleagues and the administrators what they learned by interviewing their students. It is reassuring for the rest of the faculty when they hear a veteran teacher saying such things as, "I have been teaching for 22 years. I talked to one of my 4th graders for ten minutes and I learned more about what children think and know about the concept in ten minutes than I have known for 22 years!" Or, when a new teacher shares his/her success in implementing the teaching strategy and witnessing the so-called "not-book-smart" kids exhibit high-level thinking and amazing ideas. The sharing session also provides administrators with an opportunity to see their teachers in a different light and to find out their ideas, successes, and frustrations in order to think of ways they can support the teachers.

These sharing sessions are constructive. They are not gripe sessions. The conversations deal with such statements as: This is what I observed. This is what the kids said. These are some of the assumptions I had and how what I observed related to them. These new strategies revealed to me that some kids had knowledge I would not have known they had, especially practical knowledge. Many kids who did not usually contribute started making contributions. Before this, I didn't know they knew or were interested in anything.

The sharing sessions in past WyTRIAD experiences have provided teachers an opportunity to learn that many others have similar concerns. The sessions have brought teachers from different grades and subject areas closer and have helped in establishing communication among teachers from different organizational levels.

When and where should collegial sharing occur?

It is hoped that collegial sharing will become a functional habit. Dialogue with colleagues about observations, possible inferences, problems, and successes strengthens confidence, through "rehearsal." It helps to generate solutions to problems, new ideas, and suggestions. It stimulates growth and enthusiasm.

Throughout the WyTRIAD experience, teachers have an opportunity to be engaged in both formal and informal sharing. Sharing is used at the beginning of each session, when participants share what they have done since the last session. This structured sharing opportunity is useful in translating observation into informed action. During peer coaching there is another opportunity for a specific, limited form of sharing.

Sharing outside the sessions and after completion of the in-service experience may be informal or formal. Informal sharing is spontaneous and may occur in the teachers lounge, at lunch, or other times when colleagues can share what happened in their classes. It may be with colleagues who are also participating in the program, those who are not, student teachers, and interns. Formally, teachers or the school as a whole can have sharing sessions as a part of a team meeting, grade level meeting, faculty meeting, meeting of curriculum teams at the district level, textbook or materials selection committee meeting, or other opportunity to help encourage and provide a model for other teachers who may not be involved.

Teaching for Conceptual Change Model (CCM)

The ideas of **construction of knowledge** and **teaching for conceptual change** were introduced in Chapters 1 and 3, and you should refer to those discussions now. We will review two highlights here since this is the point in the process when teachers will be **implementing the model** to help create conceptual change in their students in respect to the concepts they are teaching.

As detailed earlier (with references), several conditions are necessary to create conceptual change, in which a learner abandons or alters a previously held view, or mental construct.

- The learner must be dissatisfied with the existing view.

- The new conception must appear somewhat plausible.

- The new conception must be more attractive than the existing view.

- The new conception must have explanatory and predictive power.

The 6-stage Conceptual Change Model (CCM) is a constructivist teaching-and-learning strategy that places students in an environment that encourages them to identify and confront their own preconceptions and those of their classmates, then work toward resolution and conceptual change. It also models a collaborative, scientific approach to problem-solving. It consists of a sequence of six stages (Figure 3.1, repeated below).

Modeling

What is the value of modeling? Why is it important?

One of the things that differentiates WyTRIAD from other in-service projects is its emphasis on modeling for teachers and administrators those strategies that are new to the teachers (Figure 7.2). For example, many teachers have not done structured interviewing of their students. To convince them that this is a worthwhile process, they must have an opportunity to see the facilitator interview children one-to-one to illustrate the use of appropriate questions, wait time, questioning, follow-up questions, establishing a conducive environment, and other behaviors that are involved in the interviewing process.

In addition to hearing about new teaching strategies such the conceptual change model, for instance, they must see it and experience it. The teachers and administrators **experience** the CCM themselves as learners when the leaders go through the entire process with a science or mathematics lesson using the six steps of the CCM. At the end, the teachers and administrators share their feelings about their experience, reacting to it. Afterwards, the teachers have an opportunity to observe the facilitator using the CCM with their own students, ideally using the same lesson. After observing the lesson being implemented with their own children, the teachers discuss their observations of student reactions to the strategy.

The Teaching for Conceptual Change Model (CCM)

1. Commit to an outcome

Learners become aware of their own perceptions about a concept by thinking about it and *making and explaining their reasons for predictions*--committing to an outcome--before any activity begins.

2. Expose beliefs

Learners expose their beliefs by *sharing predictions and explanations*, initially in small groups and then with the entire class.

3. Confront beliefs

Learners confront their beliefs by *testing and discussing* in small groups what they observe from doing the activities and collecting data.

4. Accommodate the concept

Learners work to accommodate the concept by *resolving conflicts* (if any) between their initial ideas (based on the revealed preconceptions and class discussion) and their observations.

5. Extend the concept

Learners extend the concept by *trying to make connections* between what they have learned in class and other situations, including daily life.

6. Go beyond

Learners are encouraged to go beyond by *pursuing additional questions and problems* of their choice related to the concept.

Figure 3.1. This Conceptual Change Model (CCM) for teaching and learning aids learners in constructing their own knowledge by causing them to explicitly acknowledge and challenge existing understandings (preconceptions) through concrete experiences (Stepans, 1992, 1993, 1994). Further, it asks the learners to make connections, apply the target concept, and propose additional ideas and tests.

Some reasons for modeling

- To show what is to be done and how to do it.

- To demonstrate its effectiveness and allow for analysis and discussion.

- To use modeling as a vehicle to put research, discussion, philosophy, and/or rationale into practice.

- To overcome skepticism by teachers and administrators that the teaching strategy or other techniques are not practical, and/or won't work at my school, with my students. Modeling answers many questions about how the teaching strategy can work with a diversity of students in actual classes and how it can be implemented.

- The genuine use of and participation by real people to illustrate the strategy by modeling is different from simply "demonstrating" a technique or strategy in that participants are active, not just passive observers.

- The more modeling of components that is done, the more real, powerful, and effective the in-service becomes.

Figure 7.2. Values and uses of modeling instructional strategies as a part of professional development.

What is modeled in WyTRIAD?

Modeling is an essential, core activity for the WyTRIAD. Participants are not just told about things they can or should try and then sent off to try them. They have opportunities to see things modeled and then to do them themselves, including the conceptual change teaching strategy, interviewing, peer coaching, effective questioning, collaborative learning, direct instruction, facilitated discussion, planning, and lesson development.

From what we know from research, to feel comfortable with new teaching strategy, a teacher needs a variety of exposures, which may take the form of actually being taught a lesson--firsthand experience--then seeing it modeled with students. Therefore, we've found it very effective to model the new teaching strategy (CCM) with participants and administrators during an evening session, then with students the following day. There are opportunities to discuss what has been done after both sessions.

The **teacher's own classroom** is an ideal setting where a small number of observers will fit in the room without intruding on the students. However, if there are quite a few observers and the classroom is not large enough to comfortably accommodate them, the students can be brought into a larger area. In some cases, it is possible that 30-40 participants may observe a class of 25-30 students.

Modeling of classroom lessons as we do it occurs over two days. Using a concept or concepts previously identified by the teachers or some other relevant concept, the facilitator has prepared a lesson. The night before modeling the lesson with students in an actual classroom setting, however, the lesson is done with the teachers and administrators so they can experience the conceptual change model. They pose their own predictions and explanations, test them, work to resolve any conceptual conflicts they experience, and so on. They are encouraged to predict how they think the students will respond to the same lesson.

Then, when we do it the next day with the children, the participants often observe discrepancies between what happens and their own preconceptions about what they think the children know. Also, the participants often find that their own ideas about the concepts are very similar to what the children think. This causes genuine cognitive dissonance for the teachers. Just as for their students, this conflict becomes for the participants a basis for constructing new understanding.

Setting up the classroom modeling sessions

- **Arranging teacher and administrator schedules**

An evening and the following day are required for this session (Figure 7.1) so the administrator and facilitator need to communicate about logistics. Both teachers and administrators should participate in the modeled lesson and make predictions about the students' responses. Then, teachers need to be released from their teaching and duty schedule for the **full day** modeling in the school occurs in order to accommodate the full experience and perhaps to travel to different schools. It is also important for the administrators to observe modeling of the same lesson taught at different grade levels as a basis for comparison, to see how students at different grade levels react to both the teaching strategy and the same concept.

There are basically **three phases** of actual classroom modeling: pre-instructional, instructional, and post-instructional (immediate and reflective). The instructional and immediate post-instructional phases are repeated for each different class that is taught.

- **Pre-instructional phase: modeling with participants**

This phase occurs the evening before the modeling in the schools. During the pre-modeling period with teachers and administrators, the facilitator explains what he/she is going to do--introducing the concept, the lesson, activities, types of questions to ask during the lesson, materials, and so on. The role of the teachers and administrators is also explained--to observe, listen to the children make observations, and concentrate on the strengths and weaknesses of the models as the facilitator works with the students. They are to observe the students' behavior, what it is they do, their reactions to the information, any changes in their behavior, attitudes, or understanding of the concept, and how they react to the strategy.

NOTE FOR FACILITATOR: Since this is a first observation session for the teachers, you may want to keep it informal and allow teachers to make discoveries in an open-ended way, without an explicit format or instructions; or you may want teachers to note their anticipations in their journals the night before. They also may share with their colleagues what they think they will observe the next day.

- **Instructional phase: modeling with students**

The next day, in a school, the in-service facilitator teaches the conceptual change lesson, implementing the associated repertoire of teaching strategies (questioning, listening, collaboration), with a classroom of students. Meanwhile the teachers and administrators observe all that happens, listen to students, take notes, and watch the interaction between the facilitator and students. Teachers and administrators should watch without interfering in any way. They should focus on collecting data rather than participating in the lesson.

NOTE FOR FACILITATOR: You may need to be flexible to accommodate the release time the teachers have been given. There is a possibility that teachers at different levels may only be able to observe instruction to their own level of students; or, they may be able to observe all of the sessions with different students. The same lesson could be modeled with different grade levels more than once, at different times, permitting teachers and administrators to attend those they want to observe. Such flexibility may help to overcome the problem of all of the participating teachers being out of their classrooms at once, which may not be possible if a large number of teachers in a school are participating. It is a definite advantage to the teachers to observe more than one level, and this should not be considered to be optional.

- **Post-instructional phase: reflection and discussion**

At the conclusion of the lesson, the administrators should provide time for all of the participants who have observed the lesson to meet to analyze what they have just observed. The discussion focuses on immediate reactions, such as:

- What did you see?

- What are your immediate thoughts and feelings about it?

- What strengths and weaknesses were observed?

- What worked and what didn't work?

- What are your concerns about implementing this model with your own classes?

- How was what you observed different than what you anticipated?

Ideally, there also will be an opportunity to have a reflection session at the end of the day with all teachers from different groups, although this may depend on availability of substitutes and the number of teachers involved. In this way the participants can <u>actively share</u> their observations and thoughts after having had a chance to reflect a little longer.

Between-the-sessions activities

Interaction of the partners

The individual and complementary roles of the partners for the interval between Session II and Session III are itemized in the **Session-by-Session Schedule** (Figure 5.4). Teachers implement the new strategies, observe and record what they learn, and are fully supported by their administrators and the facilitator.

Specific notes for teachers

- Conduct additional interviews if necessary
- Complete plans for another CCM lesson
- Implement learning activities
- Collect data: make observations, keep track of changes in the students and your own teaching
- Make daily journal entries

Before Session III, the teachers continue their investigation of the concepts they have chosen. This can be through a continuation of the interviewing, but the main focus should shift to **teaching for conceptual change**. The teachers should use the conceptual strategy in their own classes and **keep notes** about how the students learn from it. Additionally, teachers should begin to deliberately think about (reflect) and keep track of the changes they experience **in themselves**. They should document and reflect on what is happening to their perceptions of students and their teaching philosophies.

The **journal** is a useful place for recording observations and ideas. We emphasize again that the journal should be a working companion, not a tidy, sanitized turn-in project (Chapter 6).

Specific notes for administrators

Administrators support the teachers by providing them time, materials, and opportunities to implement the new ideas. They may wish to observe the teaching strategy being implemented and assess its effectiveness on teachers and students.

Looking ahead, the administrators need to make sure videotaping equipment and plenty of tapes are ready and available for upcoming peer coaching sessions, if the teachers plan to use them. Also, share some journal entries with your colleagues or parents.

Specific notes for the facilitator

The facilitator communicates times he/she will be available to assist participants with ideas, materials, and resources they may need. Also, the facilitator needs to prepare for modeling of peer coaching, including identifying who will serve as her/his coach. Talk to the administrator about changes he/she has noticed. Share some of your journal entries with colleagues.

Chapter 8

Session III Activities and Topics

...clearly, one's best style is more a matter of personal preference than of proper pedagogical practice....

... Charles Bonnwell and James Eison, 1991

We had strong evidence that if we combined study of the rationale of a teaching strategy or curriculum with demonstrations of it, plus lots of practice and lots of feedback, almost any teacher could learn almost any approach to teaching... But it turned out there was a second stage of learning -- when the teacher would consolidate the strategy and adapt it to his or her own repertoire -- and skill alone wasn't enough to facilitate that. What was needed was companionship; especially companionship with peers.

... Bruce Joyce, 1987

About the session

The third WyTRIAD session is often a two-day session. A significant role of the **administrator** here, again, is to provide the time and flexibility for the teachers to participate in these development activities over the course of the two days. If the administrator observes the activities, as is expected, he or she is generally pleasantly surprised at the enthusiasm of both the teachers and students, as well as the learning that is often dramatically displayed by the students. The **facilitator** should see that large sheets of paper, markers, and masking tape are available for the sharing session, as before.

Since Session III generally begins three to five weeks after Session II, it is again important to spend some time at the beginning sharing and reflecting about what teachers have done and learned in that time. Another teaching for conceptual change lesson may be modeled. The idea of peer coaching and peer sharing, another of WyTRIAD's central elements, is introduced and modeled in Session III. Finally, some time is provided for closing discussion and for peer teachers to plan the lessons they will implement between Sessions III and IV.

Activities

Share results of observations

Teachers, by this point, should be quite comfortable discussing what they have learned and sharing stories about what has gone well and poorly in the interval. Again, it is suggested that the initial sharing be done in small groups, followed by the integration of the key points into the larger group discussion by a representative from each of the smaller groups.

Observations and reflections should be becoming more specific by this third session. They should not only involve changes seen in students and changes in perceptions and expectations of students, but should also encompass changes that the teachers see in themselves.

There is often interest in seeing the teaching for conceptual change strategy modeled with another concept, which can serve as a starting point for the new material of Session III. Again, teachers and administrators can be the students for the modeling exercise; or, if it seems more appropriate, a classroom of students can be used. The facilitator, when done modeling, should take time to point out which parts of the model were used the same as during Session II, and which parts were modified to be more appropriate to the new concept.

One technique that works well in Session III (especially if some of the teachers are enrolled in WyTRIAD for the second or third time) is to provide time for a team of teachers to model the strategy for their peers using their own students. The benefits for the teachers who model are clear. They not only work together in developing and implementing the strategy in a way that is directly relevant to their class, they also get the feedback of a classroom of peers. Such feedback is truly unusual in education, and placing oneself in the spotlight like this constitutes no small risk for the teachers involved.

Peer coaching modeled

There is extensive literature on peer coaching and its benefits (see References). The facilitator should define what is meant by peer coaching, and might share some of the benefits with the teachers. Having the teachers suggest benefits is another useful way to go about it. Ideas about peer coaching and sharing often flow quite naturally from a discussion about the teachers' modeling of the conceptual change model.

Peer coaching, it should be emphasized, is **not critiquing or evaluating** one another's teaching. The one who is coaching is not standing in judgment of another teacher. Well-implemented peer coaching is a positive, cooperative experience that (1) is guided by the needs of the person being coached, and (2) uses criteria agreed upon in advance.

It is most useful if the observing partner is keyed in to one aspect of the lesson that the teaching partner chooses, such as questioning techniques or gender treatment. Following the lesson, the two teachers discuss what was done and try to identify where and how improvements can be made. Suggestions should stay positive and productive, since the situation will soon be cooperatively reversed.

Develop learning activities for peer coaching experience

It is suggested that teachers team up in pairs by subject area or content and age-level of students. Each team then works together to plan a lesson on a given topic, and makes arrangements for the partners to take turns teaching this lesson in turn while the other one watches.

Topics in depth

Collecting data about the effectiveness of teaching and learning through classroom observation

By now participants have been introduced to concepts associated with classroom-based research and teachers-as-researchers, and have been gathering and reflecting upon their own data. The section that follows provides additional perspectives to build upon the experiences to this point in the WyTRIAD process.

What kind of data do I need to collect?

Suppose you have just begun to implement a new teaching philosophy and/or strategy in your class. What information would you need to collect in order to see if it is working? How can you be sure you remember your observations so you can analyze and make inferences based on specific observations rather than haphazardly remembered observations and simplistic or perhaps non-representative generalizations? What do you need to do so you will have something concrete on which to base your reflections and inferences?

What information do you need to collect? How do you want to focus your observations/data collection, and what do you want to record? Some examples of kinds of data to collect are provided in Figure 8.1. After you have data, the next step is to interpret your observations. Some questions to direct teacher reflections on their observations are provided in Figure 8.2. Also, please refer to the discussion of research in Chapters 1 and 6 and to several sample materials in the Appendices.

How do I collect data?

How do you collect and record data in an organized way, so you remember and analyze specifics rather than general recollections? How do you document changes that occur in the classroom, both changes in the way children learn and changes teachers see in themselves as they go through the teaching process?

As teachers teach, they collect data. Ideally, teachers become increasingly self-aware about their data collection and classroom observation with practice. There are both formal and informal ways of collecting data. **Informal** data is collected by mainly by tracking observations. Teachers listen to and make note of what children say and the questions they ask. Teachers observe how children interact with one another and the materials with which they are working. They also collect information from parents. Data also can be collected and recorded in a more **formal** way, as by using structured interviews and various assessments.

We are not suggesting you collect data on every child every day. Zero in on one or two students or one particular group each day and follow them, write down notes about the questions they ask, their behaviors, and so on. You need to spend about 2 weeks to get to all of the kids in a class, then go back and start the cycle again, with different questions or things to observe and investigate.

It's also not necessary to have answers to all of the questions you pose based on every student. The process should be helpful and fluid (with practice) rather than be an extra burden. As some authors have said, you are taking "snapshots" of a dynamic process, not a "videotaped" chronicle of everything that happens in your classroom.

Examples of kinds of data to collect

- How many and what types of questions are students asking?

- How are students relating to one another?

- Are students bringing any questions back to class? Where and when are they generating the questions?

- Are the students taking any questions home?

- Are they asking one another questions?

- Are they working with materials to answer their OWN questions rather than expecting questions and explanations to come from the teacher?

- Do students persist over a period of time?

- Are they willing to work through a series of steps to answer their questions?

- Are students learning content concepts?

- Do they understand? Are they grasping the concepts inherent in the activities?

- Do students seem to enjoy working with the activities?

- Do the students value what other students have to say?

- Do they see relevance in what they are doing?

- Do they see a purpose in what they are doing?

- Do they seem to be getting out of the activities the outcomes you intended? Is there a match or a mismatch? What could change this?

- Can they apply what they have learned?

- What are parents saying about what the children are doing?

- Is there any evidence that what's happening in class has any impact on the children's real world -- out of class?

- Do the students make connections between other activities? other classes? the real world?

And, perhaps the most important questions, the ones the observer must ask herself/himself in regard to each of the questions above:

- How do I know? What evidence do I have?

Figure 8.1. Sample questions to direct classroom observations and data collection.

Reflecting about your observations

INFERENCES

- Are my expectations of the students clear to me? to the students?

- Are they different than what they used to be?

- Are the experiences I'm providing appropriate and accessible?

- Are the experiences appropriate not only to the students' ability to understand but also to my expectations for the activity?

- Do the experiences meet the diverse needs of the children in my classroom?

- Do I listen more than I talk?

- Do I ask more than I say?

- Do I provide time for students to think?

- Are my questions appropriate? Specifically,
  - » Do the students understand them?
  - » Are they helpful in guiding the exploration?
  - » Do they facilitate the development of students' explanations?
  - » Or, do they steer students toward a "canned" explanation, such as the textbook version?

- Am I letting students have a larger role in deciding what happens in the classroom rather than keeping tight control myself?

- What _evidence_ do I have to support each of my inferences and conclusions above?

FEELINGS

- Am I having trouble letting go of my traditional role?

- To what do I feel resistant?

- Do I feel like I'm doing damage to the students by being less directive, doing less telling, covering fewer concepts, etc.? If so, why?

- Am I experiencing different anxieties in regard to my teaching than before?

- What do I feel unsure about?

- What do I feel good about?

Figure 8.2. Some thoughts to guide teacher reflections.

It's more difficult to observe changes in <u>ourselves</u>, for which a strong teammate and peer coach can be helpful in documenting observations about the way we teach. Teacher partners can, by mutual agreement, observe and record specific things about each other. A few examples are:

- The kinds of questions they ask

- The amount of talking they do

- The amount of listening they do

- The way they incorporate students' ideas and allow them to share their ideas with each other and the entire class

- Whether they respond differently to girls than to boys,

- Whether they allow sufficient wait time for students to give thoughtful responses

- Whether there are any "invisible children" in the class who go through the activities unacknowledged and nonengaged

It has been repeatedly mentioned by teachers who use the conceptual change strategy, for instance, that students who usually are not academically successful do much better with this approach than with traditional instruction and, in fact, have knowledge and skills to contribute. This is an example of a documented change in the classroom in response to a change in the curriculum <u>and</u> in the role and behavior of the teacher.

Where do I record my observations?

It is probably most useful to you to keep your journal on your desk in your classroom and jot observations, notes, and other things you want to remember directly into your journal. (See also Chapter 6 regarding notes and journals and the Appendices.) Your data may even be a list or a collection of lists--anything that will help you remember <u>specifics</u> and enable you to put together a more reflective entry after you have a chance to think about it and/or have time to analyze the information.

How can this recorded information be used?

Data provides you with information about your teaching, as well as the students' learning. If recorded carefully and interpreted without bias (as much as possible) it should help you to see things you might not have been aware of and suggest responses to dilemmas or problems.

Teachers also adjust their planning and instructional activities based on how the children are responding and the questions they ask. More or less time may be merited on a topic, or you may find that fundamental work needs to be done on developing those concepts that actually underlie or must precede the topic you intended to teach. For instance, teaching about <u>photosynthesis</u> is not effective unless students first have some understanding of energy and energy transfers, atoms, molecules, oxygen, and water. At the very least, the concept of photosynthesis has no relevance unless there is a sound understanding of the concept of "food" and its role in living organisms. Without some useful grasp of these fundamental underlying concepts, the overall concept of photosynthesis is just an abstraction, a chemical formula (which is yet another idea they must understand) to be memorized.

Some specific ways your recorded observations (a.k.a., "research data") can be useful are in helping you to:

- Analyze the dynamics of your own classroom

- Communicate to parents about their child's progress, changes, problems, achievement, growth, something special, new, or unique, and in asking parents to help deal with problems or encourage their child in specific ways

- Identify individualized, special interests and aptitudes

- Track the growth and development of individual children

- Recognize different ways students learn

- Look at learners differently; for example:

  » noticing that the kids who are not necessarily "book-smart" do well with certain strategies and conditions

  » that the child reluctant to share ideas in a large group has opened up in small groups

  » that a significant percentage of the students do not excel in a traditional setting or with traditional reading-writing-lecture-based instruction and activities tests

>   that what seems to be success or failure with science concepts may actually be a measure of language development or ability to learn from words (as contrasted to an ability to learn the concepts from concrete experiences and other modalities)

• Identify the diversity of learning styles in your class, identify your own teaching style, and try things to better match and adapt your style and those of your students

• Communicate with other teachers and with administrators

• Link with other subject areas, levels, projects, grade-level planning, district level, resources.

• Communicate to the child (!) observed strengths, weaknesses, and suggestions, helping the child to become more fully a partner in the teaching-and-learning process instead of the <u>object</u> of the process

• Design specific interventions. For instance, if you are developing a formal IEP (Individualized Educational Plan) for a child, you have good information on which to base your recommendations--you are not just "shooting from the hip."

Peer coaching

What is peer coaching and when do you use it?

Peer coaching occurs when two or more teachers collaborate to improve their instructional skills and strategies. It is the process by which teams of teachers regularly observe one another and provide support, companionship, constructive feedback, and assistance (Valencia and Killian, 1988). It might also be thought of as "peer sharing and caring" because of the nature of the give-and-take that occurs. All participants are both coach and teacher-being-coached. Several useful references about peer coaching are included in the References, and several **sample scenarios and formats** are included in the Appendices.

Peer coaching allows you to learn from the objective observations of your partner and to practice and perfect certain teaching strategies. It is a nonjudgmental process. In its purest sense, peer coaching is NOT an evaluation of teaching such as might be used for administrative purposes. **It is developmental and personal, not a performance indicator.**

How does it work?

Peer coaching consists of three parts: planning, implementation, and feedback. It may be most helpful to start with an example, in this case a hypothetical scenario and dialogue about preparing to teach a lesson and asking someone to help you through a peer coaching process (Figure 8.3). The example contains elements of the peer coaching process for further consideration and explanation.

- **Planning**

    Teachers plan together a particular lesson that incorporates a specific teaching strategy. They also agree on the specific aspects of the lesson being taught that will be the focus of the observation and coaching. For example, they may agree to look at the use of effective questioning, wait time, the conceptual change strategy, cooperative learning, or some other aspect for which the coach will be able to make observations (Figure 8.4).

    Also, the peer coaching team should be careful about not trying to use a single observation to make decisions about complex issues requiring a variety of data. An example of this kind of question might be, "Do I use more than one kind of assessment?" This question, which is certainly worth studying, is too broad as a guideline for the observer and requires more than one observation to decide.

    The partners develop **observation and feedback forms** to be used by the coach. Some sample forms are included in the Appendices.

- **Implementation**

    While the one teacher teaches, the other teacher sits in on the class. Using the form and items that were agreed upon, the coach writes down <u>what she/he sees and hears</u> related to the use of the particular skills and/or strategies on the form. Some appropriate types of comments indicate "yes, I observed it" or "no I didn't"; or "yes, I saw you do this." With the Figure 8.3 and 8.4 example, an appropriate observation to write down in response to the question about children contributing to the discussion might be "all of the children contributed" or "17 of the children contributed." An additional observation might be added if it seems relevant, such as "9 of the 14 children who did not contribute were boys."

    One of the subtleties of peer coaching is that the observer does <u>not</u> write down <u>inferences, opinions, or judgments</u>. Some examples of this kind of inappropriate note-taking might be: "this was a good question/poor question"; "the students seemed confused"; "the students really enjoyed the fact that you let everybody have something to say." There should be no volunteered, evaluative opinions about attitude, effectiveness, quality, or appropriateness.

140

"Chris, I'm going to spend a few days teaching a lesson on the structure of plants. The textbook coverage is uninspiring, and is mostly terminology. I want to do the best job I can, and I'd like the students to be able to use what they learn. I plan to use a constructivist instructional model.

I would really appreciate it if you could document some of my behaviors while I'm teaching at two different stages in the unit, on Monday and Friday. As a peer coach, you won't be giving me an opinion about whether I'm doing a good or poor job--how effective I am--just giving me information on which I can base my own inferences and decisions.

There are many things I need to know about as I teach (such as whether or not the students seem to be understanding, if the activities I've chosen are effective, whether or not I'm letting the students take the lead), but for this time, I'd like you to help me determine, specifically, whether or not I'm eliciting children's predictions. I've drawn up a peer coaching plan for MONDAY (Figure 8.4). Do you have any other suggestions?

Let's be clear what I'm asking you to do. You don't need to worry about hurting my feelings with your opinion because I'm not asking you to make any opinions or judgments, just to observe what happens and document it for me. So, "just the facts, ma'am." I'll resist asking you for an evaluation. I'd like to try to infer that from your data and decide where I might muke adjustments.

Then, next FRIDAY, after we've done the activities, I'd appreciate it if you could observe me again to make observations of my use of terminology when I help to synthesize the concepts about plant structure. Some questions I have in mind for Friday are: Am I consistent in use of the terms? Are the terms defined before or after they are used? Are the terms defined before or after the notion (concept) is introduced by the students? Do my definitions stand alone? How many definitions involve terms that need further defining? How many times do I define terms? What terms do the students spontaneously use during the classroom discussion? I won't ask you to make decisions and judgments, such as: Do the students understand my definitions? Am I doing a good job with the definitions? Are any of the concepts inappropriate for this class? I will use some of your observations plus other means of collecting information to help me answer these questions."

Figure 8.3 One side of a hypothetical conversation in which a teacher approaches her partner to set up a peer coaching experience.

Barbara's plan for   *Monday, September 25, 1995*
Life Science, 4th period   *Coach: Chris*

A. What I want to learn: **B. What I would like you to document:**

Are my questions clear?

1. How many times do students <u>understand</u> what I ask, compared to how many times do I have to <u>clarify</u> the questions?

Are my questions open-ended?

2. What questions do I ask that require only a yes, no, or identification-type answer?

3. What questions do I ask that the students don't respond to?

4. What questions do I ask that get interesting and/or thoughtful responses?

5. Other observations?

Does humor help evoke responses?

6. What questions or remarks make students smile?

7. What falls flat?

Have I managed to let everyone contribute to the discussion?

8. Tally which students contribute and how many times each. Also tally how many students make <u>no</u> contributions.

Do I allow enough time for the students to come up with their ideas?

9. Clock me on wait time and responses.

Do I allow students to record predictions in their own words?

10. How many times do I:
     - write what they say?
     - rephrase what they say?
     - change their meaning?
     - ignore or omit a student's idea?
     - impose predetermined categories?
     - interrupt?

Thanks, Chris!   ☺

Figure 8.4. A sample plan prepared by the teacher for the peer coaching experience described in Figure 8.3, including identification of (A) the questions the teacher wants to explore and (B) the specific data she wants her peer coach to collect.

Things to Observe	Evidence

1. Are my questions clear?
 	
How many times do I have to rephrase questions to clarify them for the students?
 	
How many times do students request clarification or comment that they don't understand the question?
	During the period I observed you --
	number of questions asked ⊥⊥⊥ ⊥⊥⊥ ⊥⊥⊥ ///
	
	number of questions students asked to clarify
	          ⊥⊥⊥ /
	teacher spontaneously restated question for students
	        ⊥⊥⊥ //

2. Are my questions open-ended? How often do I ask questions that require only an identification answer? What are some questions that I asked that seemed to get thoughtful responses from students?	Nearly half the questions you asked were identification type... examples: "What is the formal name for the leafstalk?" "What do we call the tissue that makes up the wood?" Good response to the questions and discussion about what happens to a tree when it gets hit by lightning.
3.	

Figure 8.4. A sample peer coaching format to be used by the coach while observing the peer coaching partner. The questions are generated by both partners and agreed upon in advance. This format is simple but effective.

- **Constructive Feedback**

The sharing of the feedback occurs as soon after the lesson as possible so both people have it freshly in mind. It doesn't have to happen immediately because specific information is recorded on paper. Most feedback about a carefully planned and executed peer coaching session takes 5-10 minutes, especially if it only involves sharing the observations.

Because it is based on data collected in the classroom, feedback should be accurate, specific, and nonjudgmental The observer only reports on those things that have been identified and agreed upon in advance, using the prearranged form. Objective data about the specific items are reported and discussed with the teacher being observed, using examples. For example, "I heard you asking this question, which caused students to start making predictions." or "These are the kinds of questions I saw you use that caused the students to start making predictions." Some examples of open questions you used were "What do you think will happen to the seed?" "Does anybody have any other ideas?" The teacher had asked to see if all students participated, so the observer would report "All but two students shared their predictions with the class."

Some examples of inappropriate feedback a person might request from the coach or that might be volunteered by the coach are: "Did I do a good job?" or "Is the topic appropriate?" The first question is not appropriate because it seeks a conclusion or judgment from the observer. The second question is inappropriate for peer coaching because it must be answered by the teacher, based on a variety of data. For instance, the teacher will need to make observations, ask questions, listen to students' questions and explanations, use interviews before and after the lesson has been experienced, decide how the activities are or are not helping them change their naive ideas, determine if the students see relevancy, can apply the concepts, etc. **The purpose of peer coaching is specific and different from the purposes of other kinds of strategies and information-gathering, as above.** Peer coaching can only help one small part at a time. **It cannot help with student performance, just teacher performance.**

Human nature as it is, however, it is very difficult for us to observe without making some judgments. These opinions are not to be recorded or reported. However, IF after the observations have been reported and reviewed the teacher makes a request for other information, judgments, or recommendations, THEN the observer may extend her/his comments informally. The purpose of peer coaching is to help teachers improve. As long as they are working in a trusting atmosphere, recommendations can be made after the constructive feedback is given.

Although a teacher may be shy about the process at first, teachers who develop good peer coaching relationships are enthusiastic, and note important **benefits**. They learn a variety of techniques from one another that they can integrate into their own classrooms. Some of these techniques improve discipline or classroom management. Others relate to improving effectiveness of instruction and assessment. Commonly, once teachers get into this process they want to videotape themselves in order to study their own teaching style and techniques. One of the most powerful benefits of peer coaching is the development of a trusting, collaborative collegial relationship.

What is necessary to insure successful peer coaching?

Three things are necessary for peer coaching to succeed. These conditions are a combination of individual and cooperative factors.

- First, teachers must have the necessary support from the **building administrator** to work with other colleagues during the school day. This means being given time for planning, observing, and feedback. The coach needs to be away from his/her class to observe. The administrator also must have a clear view of the purpose of peer coaching--it is to help develop best practices and skills and it is not to be used for teacher evaluation. Thus, peer coaching is not be confused with systems for peer evaluation that are sometimes used.

- Second, the **colleagues** have to be willing to share and trust each other and not be afraid to take risks in front of each other. They will work in a collaborative and respectful, safe relationship, not one that is judgmental or evaluative. Also, the confidentiality of the relationship should be respected and what the coach observes is shared only with the person being observed.

- A third factor, while not as critical as the first two, greatly enhances the peer coaching and sharing experience. It is extremely helpful if the colleagues are working at or near the same grade levels or on similar or related topics so they can **plan together** on lessons they will both use. Thus the work will be mutually beneficial and a collaborative team is formed.

Between-the-sessions activities

Interaction of the partners

The individual and complementary roles of the partners for the interval between Session III and Session IV are indicated in the **Session-by-Session Schedule** (Figure 5.4). Teachers implement the new strategies, observe and record what they learn, and are fully supported by their administrators and the facilitator.

Specific notes for the teachers

The <u>new</u> between-session assignment for each teacher is to teach a lesson with a peer coach watching, and then for the peers to share ideas based on what was seen and what has been learned from the other aspects of WyTRIAD.

All of the other components of WyTRIAD introduced so far <u>continue</u> to be integrated and studied. Teachers implement the learning activities they have developed, practice peer coaching and sharing, collect data through classroom observations, and maintain journal entries, interview students, utilize the conceptual change model, and review textbooks and other learning materials.

A note for both teachers and administrators

Teachers and administrators should bring district curriculum materials to Session IV and be prepared to review them in the light of what has been experienced and their reflections on these experiences.

Specific notes for administrators

Between Sessions III and IV, administrators work with the teachers to set up the needed schedule shifts for peer coaching and sharing, and continue their own reflective journals. The administrators' support and flexibility is <u>especially important</u> between the third and fourth sessions, as the peer coaching and sharing activities require more time and flexibility than a teacher can find in the normal schedule.

Specific notes for the facilitator

Again, the facilitator provides assistance as needed, is available for questions, and makes all needed preparations for the next session.

Chapter 9

Session IV Topics and Activities

Teaching is possibly the only profession where the customer is regarded as very nearly always wrong. Decisions are made about children and what they want without ever asking them or talking to them.

... Kate Hart, 1984

Teachers have far more freedom to innovate than they ever use. When the classroom door is closed nobody, but nobody, knows what is going on in there except the teacher and the students. ... Teachers may not be able to change the educational system, or their administrators, but the variations possible within an ordinary classroom are almost limitless.

... Arthur Comb, 1979

About the session

During the fourth session, the focus of the WyTRIAD becomes less concerned with the details of teaching, which were dealt with in the first three sessions and given over to the teachers and their professional decision-making. The focus now moves toward examination of bigger issues, including:

- Integration of disciplines

- Curricular revision

- Greater application of the teaching for conceptual change strategy to other concepts or subject areas

The tools and philosophy developed by the teachers and administrators themselves over the first several sessions and the intervening weeks serve as the basis for the step to these greater issues. **The fourth and fifth sessions are designed in part to help schools to become independent in their efforts to improve teaching.** The teachers and administrators are to have brought district curriculum materials that they have begun to review with them to initiate this process.

Before these broad, potentially overwhelming issues are introduced, however, it is again important to spend some time addressing issues and questions and sharing stories of the successes and learning experiences of the previous weeks. As before, working in groups is helpful, but this time the procedure should occur almost automatically since it is a familiar session-opening process.

Activities

Session IV begins again with participants **sharing** their experiences since the last session, particularly in regard to their classroom observations and their peer coaching and sharing. By now, the comfort level and collegiality should be well-established among participants, along with increased self-confidence about their abilities to understand what is happening in the classroom and make informed decisions about teaching strategies and curriculum. As a result, the participants are able to encourage one another as they listen and share.

Another result of the sharing is that participants now **explore and integrate** their colleagues' ideas into their own classrooms. Some time should be devoted now to this process, in which colleagues share and identify each other's specific ideas, activities, and approaches related to topics they plan to teach.

The **new material** for Session IV begins with another modeling session. Normally, this is a lesson in a different science area, a non-science area, or one that integrates science with another discipline, such as mathematics. Again, the role of students can be played by the teachers and administrators, or actual students can be used, depending on what situation is available that day. A sample mathematics CCM lesson on Area and Perimeter and an integrated CCM lesson on Levers are included in the Appendices, and may be used for the modeling. Sample earth and life science lessons are also included.

As before, each modeling session is followed by an open discussion. This discussion should focus on what was surprising and what seemed especially effective, as well as ideas for improving the lesson or making it more applicable to a particular teacher's needs.

Some time is then spent by the teachers developing new learning materials. This activity can take on the issue of integration of subject areas, or it can continue to focus on improving, adapting, and designing conceptual change lessons based on what has been learned so far.

There is some flexibility about how the remainder of the day can be spent. Choices should be made by the facilitator based on the needs and the concerns of the teachers and administrators. Ultimately, the teachers should look ahead to determine what assistance would be most useful to them as they prepare to take on the next between-session tasks. A part of the preparation for the fifth and final session is to continue with some of the observations and ways of thinking that have preceded, such as interviewing and peer sharing, and to implement some of the ideas developed from the discussion on integration.

But a key part of the assignment is to **focus on curriculum**, and the participants will be asked to look at their district's current curriculum as a starting point. One helpful tip in that regard is that it is not useful to try to study and evaluate the entire science curriculum at once! It is much less intimidating to examine a strand or even a single concept, and follow it from Kindergarten to grade 12. As for this session, teachers should bring copies of their district curriculum with them to Session V, as well as their notes, analysis, and suggestions about the part of the curriculum they have begun to examine, so there can be a useful discussion and working session.

At this time, the teachers and administrators can spend some time working in teams looking for methods of improving the curriculum based on the questions suggested later in the Topics-in-Depth section of this chapter. Emphasis should be placed on "curriculum" as a dynamic, changing process, rather than as a document that is finished, then placed neatly on a shelf.

Topics in depth

Creating curriculum that is appropriate and accessible for students

Examining curriculum

Over the course of the WyTRIAD, participants have been collecting their own data about their students and the teaching-and-learning process as it happens in their own classrooms. By now they have a basis on which to examine curriculum to determine if it matches, or "fits," the needs and capabilities of the students. It is at this point, then, that individual and district curricula need to be examined.

A focus on creating appropriate and accessible curriculum as part of the WyTRIAD is consistent with the philosophy and rationale developed from the very beginning.

- If we want meaningful change to take place it will need to be implemented by the teacher.

- Changes in the classroom should be based on a researched understanding of the teaching-and-learning process.

- If we are if talking about change, we are talking about curriculum, instruction, and assessment.

The major objective of this in-service program and the activities in which the participants engage is to provide teachers with specific skills, background knowledge, experience, and confidence to start examining what goes on in the school, and then to create appropriate changes. The experiences in the WyTRIAD prepare teachers to start looking at the curriculum in terms of their expectations of students, based on what they have learned from their own research. What they have learned from interviewing, sharing with colleagues, observing students and colleagues, working with the in-service facilitator and seeing new strategies modeled, practicing the new strategies, developing conceptual change lessons, and analyzing textbooks and other learning materials has helped the teachers to grow professionally and become informed decision-makers. **They now have a rationale for the changes they will make.**

Furthermore, all the experiences they have had are relevant to their own students because they have been implemented in their own classrooms, in their own school districts, with their own administrators and colleagues as partners. They know how to determine what is appropriate for their students. It is a natural extension of this training that they continue to learn through the research activities and that they take control of the teaching-and-learning process. They are ready to examine and make decisions about setting expectations for students, curriculum content and design, and assessment. They have learned how to make these decisions from the inside instead of the outside, and they have access to continuing assistance, educational research, and new teaching strategies through their relationship with the professional development facilitator.

By now it may be second nature to the participants to approach their evaluation of the district curriculum by **identifying questions** against which to test the curriculum (see topic below). Throughout the WyTRIAD experience, questions have been the basis for gathering information, from interviewing to peer coaching. Questions have great power and keep a process vital. Generating and answering <u>questions</u> keeps a process moving toward the future, rather than settling it into the present like <u>statements</u> do.

Even after we get to the point of stating that this is the curriculum--even after it has been revised--we **continue asking questions and we keep revising based on what our inquiry reveals**. Is this the best practice? the best we can do? What specific aspect of the curriculum can we research this semester? What kinds of assessments do we use and are they doing what we think they are doing? Are we in line with national recommendations? What topics can we justify introducing at a grade level? Can the students learn (construct) these concepts at their present level of development? For example, teachers can research such topics as:

- The appropriateness and accessibility of the concept of solar system for 1st graders

- Teaching about molecules at 4th grade

- Studying living versus nonliving and parts of the cell for 4th graders

- The use of models to teach matter and its properties

By asking research-oriented questions about the curriculum, teachers engage in examining their own expectations of students, the instructional experiences they provide to help the students meet these expectations, and the assessment strategies that correspond to the expectations they have set. In this way, all three components of the teaching-and-learning process are aligned, in both the classroom and the district curriculum.

Questions to ask when examining the degree to which curriculum, instruction, and assessment are aligned

Creating and implementing sound curriculum requires that the following questions about the concepts and approach be examined. In answering these questions, we arrive at an answer to two fundamental questions about any element of a curriculum: **Why should it be included?** and **Why is it important?**

The criteria listed in Figure 9.1 emphasize the learner as the basis of curriculum development, rather than emphasizing resources outside and independent of the learner, such as tradition and adult perspectives. When designing a learner-centered curriculum, test each component against these questions. The teacher also asks these questions of the **instructional experiences** she/he designs, not just the curriculum. As a practical matter, how effectively will the teacher be able to teach the topics and concepts within it, and how well will the learners be able to learn, to construct the concepts?

- **Is it appropriate?**

Appropriate means does it match the intellectual and developmental level of the learner. This criterion takes into account what is known about the necessity for concrete experiences, for instance, and where the learner is in the process of developing abstract reasoning abilities. **Just because a topic or concept is presently in a textbook or curriculum doesn't mean it's appropriate.** Teachers have a right and a responsibility to question whether or not something ought to be taught to the students in her/his classroom, based on his/her own inquiry-- interviewing, observing, researching.

- **Is it accessible?**

Accessible means that the student has the opportunity to make sense out of it in his/her own way by working with materials and going through experiences. For something to make sense, the learner needs to be able to construct a concept rather than just mimic and play back what the teacher, book, vocabulary list, or problem sheet has said. Also, is the concept teachable in this setting, with these students? Can it be discussed meaningful in language that is understandable to the student, and are the necessary underlying concepts in place?

**When judging if something should
be included in the curriculum, ask**

- Is it appropriate?

- Is it accessible?

- Is it meaningful?

- Is it relevant?

- Can it be taught effectively?

- Can it be assessed effectively?

- Does it provide students opportunities to raise their own questions?

- Does it provide students the opportunity to want to learn more about the concept?

- Can students see the relevancy and applicability to real-life situations?

- Does it make connections

  » between concepts?

  » between disciplines?

  » between school and students' experiences outside school?

Figure 9.1. Questions to help critically examine whether or not an element of the curriculum should be included, as well as to aid in assessing the degree to which curriculum, instruction, and assessment are aligned.

- **Is it meaningful?**

A meaningful experience, topic, or concept has some personal **value**. Is it important to learn it? Is it interesting? Does the learner want to learn it? Is there a reason for the learner to want to learn it? Learners are more likely to respond to something that makes a personal connection of some kind, such as explaining an observation, meeting a need, or answering a question they have. Is it a central concept or a trivial "factoid?" Is it something helpful, information students can use to solve problems or reason further? Is it challenging? Does it go beyond what is already known, build on it, and form a basis for future building? Does it reward mastery, or is it unnecessarily repetitious? Something need not meet all of these criteria to be justified, but it should meet some of them.

- **Is it relevant?**

Relevancy means that students see some application to real-life situations or a connection to existing knowledge and/or to other aspects of experience.

- **Is it challenging?**

Something that is challenging causes the learners to use imagination and what they already know to **expand knowledge**. It stretches or confronts an existing belief or poses a problem. It does not mean making something hard to grasp, exceptionally frustrating, or confusing. It does not mean just providing "more to do." The level of challenge should be appropriate to the level of the learner and lead to growth.

- **Can it be taught effectively?**

Teachers implement curriculum, and need to be able to provide instructional experiences that help students to develop an understanding of the concept. Does the item allow the teacher to provide students opportunities to learn that are interesting, challenging, and relevant, that include tangible experiences for concrete learners, and that can extend and apply the concept. Does it permit the teacher to use strategies beyond lecturing and memorization?

- **Can it be assessed effectively?**

If we set expectations for learners that deal with different domains (*e.g.,* the student's ability to apply information and raise questions, willingness to be involved, skills, attitudes toward learning) and that acknowledge different ways of knowing and being able to do, we must be able to design ways to assess, evaluate, judge, determine, or otherwise measure these expectations. Can we legitimately tell if instruction has had the desired impact on students--a genuine difference, not just superficial or contrived artifact?

Assessment reflects not only the performance of the learner, but also the effectiveness of the process. We often miss this point. (See also Chapter 1.) Assessment practices need to be continuously challenged, evaluated, and brought into line with expectations and instruction.

- **Does it provide students the opportunity to raise their own questions?**

    The person who formulates the questions is the person who is doing the thinking. If the teacher is the only one asking questions, the teacher may be the only one who is actively thinking--about the information, its application, ways to demonstrate relevance and make connections, and the direction of pursuit. Knowing they are expected to ask questions can invoke students' own sense of wondering about something. Questions reveal gaps in the constructive process, and formulating them clarifies what learners do and do not know, makes it more explicit. Asking the questions--instead of just getting answers to questions the teacher poses--empowers the learner to be the learner. Teachers need to not squash or rush questions, but respect the right and foster the opportunity for students to ask them, to have the opportunity to learn the things that are important and meaningful to them

    A common, unacknowledged gap in curricula is in not allowing young learners opportunity to learn <u>how</u> to formulate questions--as a part of their language development and as a learning strategy. They also must learn how to formulate the kind of questions that can be investigated via science.

- **Does it provide students the opportunity to want to learn more about the concept?**

    To meet this criterion, something must open new possibilities, keep the learning process open-ended.

- **Does it make connections?**

    Information is valuable and comprehensible when it is **connected** to an existing mental framework. Often it takes a long time for a learner to have enough related parts of the framework--information and experiences--to construct the connecting linkages. Traditional curriculum and instruction can, at its worst, be a dump-truck approach that attempts to fill the learner with shovelfuls of fragmented "input." On the other hand, coherent and cohesive curriculum and instruction will deliberately assist learners in making sense of it all. By making connections we mean between concepts, between disciplines, and between real life experiences and things that are learned at school.

Between-the-sessions activities

Interaction of the partners

The individual and complementary roles of the partners for the interval between Session IV and Session V are indicated in the **Session-by-Session Schedule** (Figure 5.4). Again, some of the responsibilities between sessions are similar. Administrators and the facilitator continue to assist the teachers as needed. **All** parties maintain their journals, recording their observations and subsequent reflections.

Specific notes for the teachers

- Continue observations
- Implement new integrated lesson
- Explore the "fit" between students and curriculum
- Daily journal entries
- Make daily journal entries

The **teachers**, as education professionals, implement the teaching materials developed, continue to collect data, and review their own use of curriculum, instruction, and assessment and look for ways to make it more effective.

Specific notes for the administrators

The administrators support the teachers by providing time and flexible scheduling as needed for peer planning and interaction. Administrators also locate and assist with curriculum documents and participate in examining the curriculum.

Specific notes for the facilitator

The facilitator remains available to answer questions and assist in any way needed. Also, the facilitator prepares to <u>evaluate</u> the effectiveness of the process.

Chapter 10

Session V Activities and Topics

What is different about the current movement [toward standards-based educational reform] ... is that standards should be made explicit and should center on learning. Up to now, our standards have been implicit--they have had to be inferred from the content of textbooks, course syllabi, examinations, and state and district curriculum documents. They have been more concerned with action than with results, more focused on what to teach and ways of teaching than with what is to be learned as a consequence of teaching.

... James Rutherford, 1995

About the session

Session V is the last WyTRIAD session--at least for some districts. Other districts seem to resonate with the WyTRIAD structure, and choose to re-enroll the following year. In one case, a district chose to be involved three years in a row, then continued the process on their own, using local expertise gained from WyTRIAD to replace the university researcher in the role of facilitator. Clearly, the goals of WyTRIAD--research-based examination of teaching and learning, considering preconceptions and targeting conceptual understanding, and cooperation--are goals that can help the school that chooses to work with them.

As with Session IV, Session V takes a look at the bigger picture. This time the picture goes beyond issues of curriculum and assessment in the school or the district, and looks at these issues as they have been examined nationally in the last few years. National trends and events such as the publication of the

National Council of Teachers of Mathematics *NCTM Mathematics Standards*, the AAAS *Project 2061: Science for All Americans*, and Benchmarks *for Science Literacy*, and the National Research Council (NRC) *Everybody Counts* and *National Standards for Science Education* serve as a starting point, then the discussion returns to the local school. "How can these documents be used as resources to help us revise and improve our local curriculum?" is a question that serves to unify the fifth session. This discussion begins just after the usual sharing at the beginning of the session.

Activities

At the beginning of Session V, participants share the results of their classroom observations, continued peer coaching, and implementation of their integrated lessons. They also share their observations and conclusions about reviewing the district curriculum, particularly in regard to ways the curriculum does or does not fit with what they have learned from their own classrooms during the project.

The in-service facilitator then conducts a discussion of the impact national trends and resources can have at the district level. After attention to any additional issues the teachers and administrators may wish to discuss, some time is devoted to looking toward continuing and expanding the ideas and strategies that have been implemented. Finally, there is a discussion regarding evaluation of the in-service experience.

What have all participants learned from participation in the WyTRIAD? Session I began with what educators in general have learned from the research on teaching and learning. In Session I the partner from outside came in to share experiences and the rationale for looking at different ways of teaching. Now Session V focuses on what each participant has learned personally, and what the participants can do themselves.

This session brings the process to full circle. It gives teachers ownership of the process. They ask the questions themselves to continue it. They take ownership of the curriculum. They take leadership.

Topics in depth

How can our expectations of students, the instructional experiences we provide, and assessment be aligned?

It may seem obvious that instructional experiences and assessment should be geared toward achievement of meaningful expectations that we set for student attainment. Such is not always the case, however.

Expectations

If we think our students ought to develop a conceptual **understanding**, patterns of **thinking**, positive **attitudes**, and collaborative **skills**, then we should set our expectations keeping these in mind.

Our increased understanding about the teaching-and-learning process and national trends is changing our expectations of students. The priorities are shifting from a high emphasis on a factual knowledge level to an emphasis on transforming the student--incorporating conceptual understanding, ways of knowing, learning how to learn, and developing positive attitudes and values toward the world around them. There is a **shift** in emphasis, for instance, from how many terms a student can get right on a test to how well the student does at reasoning, problem-solving, and applying concepts.

The work of the NCTM led the way in revamping the way mathematics education should be viewed, and provided a model for a national reexamination of science education. Both Project 2061's *Benchmarks for Science Literacy* and the NRC's forthcoming *National Science Education Standards* (NSES) are designed to provide a solid conceptual basis for K-12 science education reform. The work of thousands of educators and science and technology experts has resulted in recommendations that delete some traditional content and materials which do not directly meet specific learning goals. Missing from both documents, for example, are simple machines, plant and animal phyla, and balancing chemical reactions. **The emphasis is on identifying central concepts that children at each level are capable of understanding and that can be developed progressively and coherently throughout their K-12 experience.**

The issue of connections and integration of subject areas is central to standards-based reform, but presents some interesting challenges for teachers and other curriculum developers. Project 2061 is developing a computer-based tool,

Resources for Science Literacy: Professional Development, to cross-link Benchmarks, NSES, and standards documents developed by the NCTM and National Council of Social Studies (NCSS).

We can evaluate local curriculum in response to these national trends and the movement away from a nearly complete emphasis on recall of factual knowledge in assessment toward more diverse measures of learning and knowledge by asking some questions when developing the <u>knowledge portion of the science curriculum</u> (Figure 10.1).

Choosing the knowledge base of the curriculum

- Why should it be included?

- Why is it important?

- Does it motivate the learner to raise significant questions?

- Is it applicable to everyday experiences?

- Does it incorporate meaningful themes in an integrated way?

- Does it utilize science processes meaningfully?

- Is it connected to other areas of science or to other disciplines?

- Does it address attitudes or habits of mind?

- Is it accessible and appropriate for the learner intended?

- Does it help the learner to construct meaning?

- Can it be taught meaningfully?

- Can it be assessed effectively?

- Is it part of the entire spiral? Are the steps manageable for the learner?

- Does the learner have the preconcept, skill, and intellectual development necessary for the concept?

Figure 10.1. Questions to ask when deciding on what concepts to include in the knowledge portion of the science curriculum.

We are challenged to develop expectations that are in line with these directions. The expectations are different in tone and scope than traditional content-oriented objectives. For example, Figure 10.2 lists some of the expectations we may have of our students as they study a particular concept, whether it be pendulums, levers, plant growth, or perimeter and area.

As they study [concept] ...

1.   **Students will become aware of their beliefs related to [concept].**

2.   **Students will show respect for the opinions of others.**

3.   **Students will identify and control appropriate variables.**

4.   **Students will communicate and analyze data using various modes; for example, speaking, writing, tables, graphs, drawings, photographs, collections, computer programs.**

5.   **Students will go beyond with their knowledge of [concept] outside the class, continuing to test their understanding.**

6.   **Students will enjoy being challenged.**

7.   **Students will develop a conceptual understanding of [concept].**

Figure 10.2. A set of seven sample expectations for students that are appropriate for guiding the development of instructional experiences and assessment tools.

Having chosen these particular 7 **expectations** for the students (presumably based on our determination that the target concepts were reasonable and appropriate), we then have the responsibility to provide them with appropriate **experiences**. The conceptual change learning model is a vehicle for doing this. Finally, the **assessments** we plan need to follow through with this sequence.

Experiences

Using the 7 expectations listed above, what **experiences** would we need to provide so students can meet these expectations? To know where to start with the concept, individual interviews are important. Figure 10.3 suggests ideas for experiences that are <u>keyed to</u> the numbers of the expectations in Figure 10.2.

Developing EXPERIENCES to match EXPECTATIONS
in regard to the concept of seed germination

1.   To help students become aware of their own beliefs about seeds, you might create an opportunity for them to talk about seeds, ask a question, or look at a variety of seeds (such as those of peanut, mung bean, radish, alfalfa) to stimulate them to make statements about seeds. Be alert to helping students distinguish between what is actually part of the fruit and the seed it encloses, to avoid creating or reinforcing misconceptions.

2.   To help students meet expectation #2 they need to be placed in groups with all students having an opportunity to verbalize and give opinions about seeds or the experiment about seeds. For them to feel comfortable about exposing their ideas, they must behave in a way that provides respect for all opinions.

3.   Provide experiences to help students identify and manipulate variables having to do with the germination of seeds, such as light, temperature, moisture, and soil. Plan how you would help students develop appropriate hypotheses and design experiments to test them. Provide two or three kinds of seeds that sprout rapidly and readily in either soil or a hydroponic environment, such as radish, mung bean, alfalfa, or Wisconsin FastPlants™.

4.   Students decide on the kinds of measurements they will take, at what intervals, then present the data in a systematic way using tables, charts, or graphs. They also may consider using drawings, photographs, and xerographic images to aid their analysis. They should be able to explain what they did to other students, and make some conclusion in regard to their initial predictions and explanations. This topic lends itself to preparation of a meaningful laboratory report, or inclusion in a portfolio.

5.   Encourage students to look for seeds in other places at home and outdoors. They may become interested in dispersal adaptations of seeds or seeds that are used for food and seasoning. Help them to find other questions, problems, and activities on this or related topics.

6.   As a research subject for students, seed germination is easy, cheap, offers many examples for making connections, and lends itself to integration with mathematics and social studies. Designing ways to test students' predictions should be fairly straightforward, and lend themselves to some creative "what if" extensions.

7.   In the experiences we provide, students have to have the opportunity to confront their own ideas by expressing, testing, resolving any conflicts, reading, and listening to each other to develop a conceptual experience. So, the

combination of interacting with materials and each other should help each student construct a viable meaning of the concept of seed germination in her/his own way.

Figure 10.3. Seven sample experiences aligned with and designed to meet the seven expectations of students that are identified in Figure 10.2.

Assessment

Since assessment is an important part of the learning process, it should be tied closely to (1) our **expectations** of the students and (2) the **experiences** we provide for them. From reviewing the sample list of 7 expectations in Figure 10.2, it is obvious that we need a variety of ways to assess whether or not the expectations are met. In addition to teacher assessment, we should provide students with opportunities to evaluate themselves with respect to their own ideas and changes in their skill levels, attitudes, and behaviors.

To gain insight into students' conceptual understandings, skills, attitudes, and behaviors, we need to use **various strategies** to collect necessary information. Some of the strategies are:

- Interviewing

- Observation

- Performance tasks

- Pencil-and-paper tests

- Written reports

- Projects

- Journals

- Portfolios

While it may be easy to list assessment strategies, it may be a little more challenging to decide how to apply them. Listed below are six learning situations for which assessment **examples** are provided in the Appendices.

- Using science processes and laboratory activities

- Working as a member of a cooperative group

163

- Working as a member of a jigsaw-formatted group

- Making concept maps

- Reflecting on experiences and thought processes

- Working on an investigation

Example: Pendulums (Stepans, 1994)

The sample lesson on <u>Pendulums</u> was provided to be modeled during Session II. The teaching for conceptual change strategy was used for the several activities that make up the lesson. These experiences were based on expectations, such as those listed in Figure 10.2, although there were more of them (18). The lesson had an **assessment component that was specifically designed to measure the degree to which students met the expectations**. A sample set of potential expectations and how they are matched to instructional experiences and appropriate assessments is included in the Appendices. The full assessment chart is also presented in the Appendices. One example, however, is presented in Figure 10.4. Note that this format for planning assessment--really for aligning the entire process--also would work for the seeds example, or any other curriculum topic.

How to share and extend your experiences

The following list represents some of the ways past WyTRIAD participants have been able to share, disseminate, and extend their experiences through speaking, writing, and consulting:

- Help other colleagues in their schools

- Conduct district in-service

- Work with ongoing revision of district curriculum

- Make presentations regarding the project to school board, parents, and community organizations

- Share the process and information gained through the project at state, regional, and national meetings

- Prepare publications and correspondence

In addition, a high percentage of WyTRIAD participants have decided to pursue advanced degrees.

Aligning EXPECTATIONS with EXPERIENCES and ASSESSMENT

Expectation #1 from a list of 18: "Students will become aware of their beliefs related to the behavior of the pendulum."

Experience related to this expectation:

Situation Presented to Student	Suppose you have two pendulums, equal in size, and one has a wooden bob while the other one has a metal bob. Predict what would happen if you pulled both of them toward you and released them at the same time. Provide reasons for your predictions.

Assessment related to this expectation:

Assessment Strategies Used	Observations
	Interview (individual or group)
	Performance

Expectations Targeted	#1

Assessment criteria:

Expectations Met	Willing to make prediction
	Able to predict
	Willing to write down reason
	Able to come up with sound explanation

Expectations Developing	Attempts to predict
	Makes effort at reasoning
	Attempts explanation

Expectations Not Met	Resists making prediction
	Makes no predictions
	Resists writing reason
	Unable to explain

Figure 10.4. An example of how expectations for students, instructional experiences, and assessments are brought into alignment to provide a cohesive, carefully designed teaching-and-learning process. This example is based on the Pendulums lessons in the Appendices.

Looking ahead

During Session V, members of the partnership discuss things they could and would like to do to continue the process they have begun. Their ideas have included:

- Find ways to extend and build upon the partnership.

- Plan how to involve other schools and colleagues.

- Decide how they can help other schools to get started, by sharing their expertise and skills.

- Continue to develop more units of instruction for their own classrooms.

- Share units with other teachers on grade level who have not participated.

- Revise science curriculum for the entire grade level.

- Develop appropriate ways to assess the effectiveness of the new strategies and approaches for both teachers and students.

- Extend the model to other subject areas.

- Propose restructuring of science curriculum across all grades in the school.

- Propose restructuring the teaching environment school-wide.

- Disseminate the same sort of information outside the school.

- Continue to develop teaming and peer coaching.

- Continue to strengthen all components of the experience.

- Eventually assume leadership role in conducting similar projects with other teachers.

- Identify within own ranks future leaders to conduct similar projects to expand the approach.

Here is a good example with which to end this section. Participants (teachers and administrators) from one school district conducted a panel discussion where they talked about their experiences, strengths and weaknesses, and what they have learned. They did sample activities with teachers and administrators from other schools, the district central office, the state department of education, people from universities and science education centers, and parents. They drew conclusions from what they had experienced and made suggestions for sweeping changes at the school and district level. The changes they are proposing are based on what they have learned from their students and how they have started looking at curriculum, the teaching-and-learning process, students, and themselves differently as result of these experiences.

PART THREE:
Evaluation, Barriers, and Continuing the Process

Chapter 11

How effective is the WyTRIAD?

The result of successful involvement of all stakeholders is "systemic change" -- real, coordinated, effective change in the way schools work to educate children.

... Joseph I. Stepans, Barbara W. Saigo, and Christine Ebert, 1995

What are participants saying about WyTRIAD?

Many of the WyTRIAD teachers have been interviewed over the past three years to determine their responses to the program and to evaluate it. Most recently 19 teachers who were participants during the 1993-94 school year were interviewed as part of a WyTRIAD evaluation (Kleinsasser and Miller, 1994). Most of the teachers taught in grades K-8, and two of them taught in high school. The set of interview questions is included in the Appendices.

Five major themes were identified in the teachers' interview responses and journal entries.

- Rethinking the learning process, particularly in response to interviewing students about concepts, and making the process more "child-centered"

- Being in meaningful conversation with children and youth about learning

- Developing professional relationships, particularly in the development of peer teams that could work together on a regular basis

- Reconsidering assessment procedures

- Building professional confidence through classroom research, enabling the teachers to make curriculum decisions not only in their classrooms, but also at the district level

Interviewing students was noted by several teachers as being time-consuming and challenging, but all reported that it was one of the most exciting, startling, and ultimately useful components, as revealed in the following quotes. Interviewing also became an important assessment tool.

"I was amazed that the kids didn't know some of the very things I think are basic and have taught in my class. The misconceptions that students have are really there."

"This interview process is [a] valuable tool in assessing entry level skills and understandings. It always amazes me how children can come up with logical reasons for common events. Their answers make so much sense even though they may not be what we are looking for. These misconceptions can color the learning process and make further conceptualization more difficult. This is where the teacher receives invaluable information on which to begin instruction."

"It helps me realize I'm developing lessons for kids, not to satisfy curriculum requirements. The kids and I plan together."

"Interviews create motivation. The best teacher is the student. Learning is a complex process and we need to convince parents, school boards, and administrators that we are professional and that with the time we've spent observing and interviewing students we know that they need time to think about what would happen, to share ideas to test and to make the connection with the expert guidance and direction of a teacher. Kids will feel better about themselves. They will have improved attitudes using the Stepans' model."

Peer coaching was also identified as a highly successful component, especially among teachers who had been using the conceptual change teaching model for two or three years, and who had been through more than one WyTRIAD in-service. These teachers worked as professional teams for preparing interview

172

questions, interviewing, and peer coaching of each other's lessons.   As one teacher put it:

> "What would really be neat would be to be able to make peer coaching as much a part of the curriculum as reading or math. We peer coached twice this year. Because of time, we took a couple of short cuts, but we did go over lesson plans, teach and have a follow up. It was great for both of us. It is refreshing to go to another room and just see different ideas and techniques. In my own room all day, I never see another method."

Modeling of the teaching strategies by the instructor was cited as extremely useful, by allowing the teachers to visualize how it is done and see it done with students. The modeling took away some of the anxiety of trying out a new strategy. As one teacher put it, "I felt more confident trying this in my classroom."

Getting teachers to think like researchers is one of the goals of WyTRIAD, but the teachers don't uniformly come out of the process thinking of themselves in this way. Perhaps the term "researcher" is problematic because it sounds so formal and remote, and, paradoxically, because it sounds so powerful. Only one teacher actually stated, "I am a researcher." This teacher (grade 4) and some of her colleagues have had strong influence on the development of the district curriculum, noting that district personnel and the upper grade teachers were impressed that they really knew what they were talking about and had information (data) to back it up. Yet, whether or not the teachers accepted the designation, it was the interpretation of several readers of the teachers' responses that they were indeed conducting research. As one reader stated, "When teachers interview students, use results of their interviews, apply new teaching strategies, collect data, and observe changes, then they are researchers."

What has evaluation revealed?

Kleinsasser and Miller drew six conclusions and recommendations from their study.

> 1.   The WyTRIAD "is a successful, teacher-friendly in-service model that challenges teachers to rethink science and math instruction in their individual classrooms."
>
> 2.   The teaching for conceptual change model "has the potential to help teachers conceptualize and take ownership

of district-wide science and math curriculum goals." It also helps teachers to "feel more comfortable and more confident in teaching science and math" and "empowers some elementary teachers who may feel less knowledgeable and less valued than secondary science and math teachers."

3.   "Interviewing K-12 learners about their conceptions and misconceptions provided compelling evidence to teachers that their instruction was <u>neither</u> as effective <u>nor</u> as productive as they thought." Interviewing brought the teachers into "real communication" with their students.

4.   Peer-coaching "produced supportive professional and personal relationships which teachers highly valued at three levels: a friendly collegial level, an opportunity to change one's teaching approach, and a strong positive relationship with the university...."

5.   Administrative support is essential for success. Since administrative support will vary, "it is likely that districts wanting to implement a teaching for conceptual change model via the TRIAD will need to evaluate the level of support needed for the project to be successful within the district as well as by each attendance center."

6.   It is recommended that the relationship of the WyTRIAD model and teaching for conceptual change be consistently clarified, to avoid confusion.

We would like to emphasize the distinction called for in item #6. **WyTRIAD is an in-service model** that engages its partners, over an extended period of time, in activities that promote a constructivist approach to both teaching and learning. **Teaching for conceptual change is a central idea** that is based on a constructivist view of learning and on which the process is focused. The **Conceptual Change Model (CCM) is an instructional model** that fosters conceptual change.

Chapter 12

Overcoming barriers

> *Having our administrator involved meant:*
>
> *... the support of feeling comfortable enough to take risks and try what might work in your classroom ... knowing that if your administrator walks into your classroom at that time they're saying, "I understand."*
>
> *... if they can't explain to a parent why this program's better, then we're out there by ourselves.*
>
> *And it was not forced ... I think that's probably why it's successful.*
>
> *They need to know what we're doing. They need to understand it, too, so that when they come in and evaluate us ...*
>
> *It was good for [district superintendent] to be there because [he] underwent some changes and without that, the curriculum wouldn't have changed either.*
>
> *...Wyoming WyTRIAD teachers, 1993*

Are there certain barriers that keep changes from being implemented?

What teachers were asked

Another set of interviews examined the experience of Goshen County schools with the WyTRIAD in a different manner (Stepans and Saigo, 1993). The 30 teachers representing 5 schools were interviewed in small groups, by schools, rather than individually. The 3 administrators were interviewed separately from the teachers. Most of the teachers had been involved with the WyTRIAD more than once. A list of the questions that were asked is included in the Appendices.

The interviews focused directly on **components of the WyTRIAD model** and the **process of change** in which the participants were engaged--change in themselves, their classrooms, schools, and districts. The interviews explored teachers perceptions of the:

- Value and importance of the partnership

- Value of the teaching for conceptual change model (CCM)

- Changes they saw in their students and in themselves

- Components of the model they were comfortable with and continued to use over a long period of time

- Barriers that prevented or discouraged them from using various components

Questions dealing with <u>implementation</u> and <u>barriers to implementation</u> of components of the WyTRIAD model were:

- Which components of the model were you comfortable with and have you continued to use over a long period of time?

- What are your perceptions of barriers that prevent or discourage you from using various components?

- What are your recommendations about overcoming these barriers in order to implement the components?

What the teachers said about barriers

There was a striking difference between the teachers who had strong administrative support for their participation and efforts and a group of teachers who did not. Special comment will be made later about this group. It was clear in the comments made by the teachers in both groups that they believed it was **essential for the principal in a school to buy into the process and actively understand and support the teachers.**

Lack of administrator support was identified as an obstacle to both the overall change-from-within **process** and implementation of specific **components**. The most practical reason was because the building principal controls the schedule. "Time" was the most frequently cited barrier.

Given that the role of the administrator is a critical underlying factor that has multiple impacts, the **barriers** identified by the teachers are listed below. Some of them are interdependent.

- Insufficient time during the day, as part of the professional assignment, to

  » Interview students

  » Collaborate with colleagues in planning, lesson development, peer coaching, sharing

  » Assess students

  » Prepare

  » Think

- Lack of administrator involvement

- Lack of confidence in their own knowledge of science and concern that they may inadvertently lead students into misconceptions

- Existing curricula

  » District criteria that include lists of concepts to be covered, whether or not the teachers felt their students were capable of mastering them

  » Lack of adequate supporting information and sample lessons

  » Textbooks

- Unavailability of resources, including materials for activities and teacher reference materials with good background information about science concepts

- Too many students in a class

- Requirements, expectations, and teachers at the next higher level driving what a teacher can, can't, and should do

- Funding, especially as it relates to

  » More demands on teacher time (because of loss of substitute teachers and aides, less flexible scheduling)

  » Fewer opportunities to participate in professional activities (because of loss of professional days and substitute pay)

  » Decrease in classroom resources (books, hands-on materials, funds for field trips, other)

  » Increased class size (because of fewer teachers hired)

- Some parents, particularly those who were very traditional and/or who were strongly concerned about test scores

- Testing, particularly district achievement tests and standardized tests (including the ACT)

- Absence of new models, strategies, and modeling in pre-service teacher education programs

- Reduced involvement by the university personnel at the end of the in-service experience

---

What the teachers said about overcoming barriers

As a counterpoint to the discussion about barriers, teachers were asked to share ideas about what they think could help them to overcome the barriers. We believe their responses are useful not only for this in-service model, but for any situation in which teachers are involved in constructive change and restructuring in schools.

- **Administrative support**

The teachers agreed that the first requirement--the prerequisite--for overcoming barriers to their ability to implement new ideas, strategies, and decisions was an actively participating and strongly supportive administrator.

- **Release time during the school day**

Time during the day could be provided by the effective use of added personnel, including a combination of floating substitutes, aides, college students, and volunteers (including parents). Some departmentalization also would help, reducing the number of individual daily preparations for each teacher.

The teachers discussed several ways additional personnel could be useful. Substitute teachers could take classes when a teacher is peer coaching another teacher or when partners are team teaching. Teacher aides and college students could assist in logistics, preparation, clean-up, gathering resources, work with a group of students while the teacher works with another group, and perhaps to manage the class while the teacher conducts individual interviews and assessments. Relieving the classroom teacher from such tasks as playground, cafeteria, and hall duty would also make additional time available.

- **Restructuring of the school day**

Recommended changes to the schedule would expand preparation and planning time and establish a regular team meeting time for collaboration.

- **Development of strong teacher teams**

Working regularly with one or more other teachers enhances planning, sharing, peer coaching, curriculum development, and other professional activities.

- **Reduced class size**

Smaller classes permit greater interaction with each individual student. The teachers felt it was especially important for investigative and group work, individualized instruction, and assessment. A class size of 18-20 students would be ideal.

- **Continuation of the partnership**

The teachers valued the continuing, intensive involvement of university personnel and other professional development resources to assist the teachers with subject content, strategies, modeling, and coaching. They felt the continued involvement of this partner help to provide needed support, visibility and encouragement, continued growth, and trouble-shooting. Also, they felt the continued support for the administrators would help maintain interest and assist in implementation.

What goes wrong without strong involvement and support from the administrator

The fundamental strength of this model for continuing professional development comes from the **partnership**. The role of each partner must be fulfilled as agreed. **This is not just "teacher" in-service, but a process that is geared to changing the school itself, from within.** Teachers who had strong cooperation, participation, and support from their principals developed confidence and success. The teachers became leaders in the change process.

The group of elementary teachers whose principal did not participate and assist them became discouraged, disappointed, and dispirited. They brought talent, experience, and ideas into the partnership, but were unable to fulfill all of their expectations. They were not given time to interview students, collaborate with each other, and think about the teaching-and-learning process. They did not have necessary manipulatives and other materials to do hands-on activities. They were reluctant to risk attempting new strategies because they were afraid the principal would not help them answer parents' questions, or understand what they were doing if he should visit their classrooms.

The teachers also felt powerless and disenfranchised, because curriculum and testing decisions were made by teachers at the junior high and high school level, then passed down to them to implement. Thus, the teachers' ability to design and implement curriculum and assessment based on their knowledge of their own students was fully preempted.

In spite of these problems, however, there was hope that they might have a better chance in the future. One teacher, in particular, was optimistic that teachers can provide leadership, given the support of their administrators and more emphasis on "the developmental aspects of education."

Chapter 13

How can we start a WyTRIAD project?

A small group of learners does benefit from the traditional approach of being presented the product of scientists' work and imagination in the form of rules, laws, and generalizations. Such an approach is not helpful for the majority of students, however. It creates in them a feeling of helplessness, forces them to merely memorize the definitions, rules, laws, and formulas and makes them grateful that it will all "go away'"as soon as the exams are over. This kind of science experience is an injustice. The world and the things in it belong to all of us, its phenomena affect us all, and we are all entitled to understand it.

... Joseph I. Stepans, 1994

Who should initiate the process, how, and when?

The first WyTRIAD projects were initiated by university personnel. Some subsequent projects have been initiated by university science education faculty, by a school principal, an assistant school district superintendent, a regional coordinator of a National Science Foundation State Systemic Initiative project, and by a teacher who is head of a high school science department.

In places where the approach has been continued over the past few years, the initiative for maintaining and expanding it to other teachers and other schools resides in the local teachers and administrators. Thus, no one member of the partnership truly "owns" it or is singularly responsible for finding out about and starting the project. Leadership for initiating it can come from any of the partners--teacher, principal, or professional development facilitator.

By the fact that you are using this book, you have access to persons who can help start a project. Because it is the most complete compendium about the model and all of its components, it would be a good idea to read chapters 1-5 in advance of an information meeting and to have a copy or two on hand for people to look at during the meeting.

- Plan a meeting. Send notices to teachers, administrators, and other interested parties to talk about the project, its components, expected outcomes, how it has worked in different places, and so on.

- Before the meeting, look at the possibility of offering course credit for teachers and administrators through a nearby university or college, or state department of education. This option is often an important additional incentive.

- Think about schedule options. WyTRIAD projects usually take place over a semester (or quarter). Propose tentative schedules that correspond to fall and/or spring semesters.

- In addition to the professional development facilitator, teachers and administrators who have already implemented the program might speak about the impact it has had on them, their schools, and their students.

- At the meeting, explicitly go over what will be expected of all partners in regard to time commitment and cost for both the sessions and the activities between the sessions.

- After the expectations have been communicated, ask if people are interested in signing up for university credit and initiate the sign-up.

Who should be involved and in what roles?

All stakeholders in the project should be involved in the initial information sessions and preliminary planning, including the district superintendent and other appropriate district personnel (such as staff development coordinator,

science/mathematics supervisor, curriculum supervisor) and **school** personnel--
the principal and appropriate coordinators.

As many **teachers** and **administrators** as possible should be invited, not just delegates or designated individuals representing a category, school, or office. Everyone who is a potential participant should attend the initial meetings.

Parents also can be invited to the initial meeting or to a later meeting at which they can participate in a learning activity using the conceptual change model. This usually generates quite a bit of excitement. Although parents will not be involved in the full activities, they are important partners in understanding and supporting the project; thus, they should be kept informed and be given opportunities to experience some of the things that happen in the classroom. Their children are the primary beneficiaries of the process, through better and more appropriate learning experiences and assessments. The enthusiasm of children who learn in such active classrooms should be communicated to and supported by the parents whenever possible.

All potential participants should be willing to commit to the entire **4 to 6 months** of the project. Most critically, for a school to participate, there must be a commitment from the <u>principal</u> to both be present at each session and to support and assist the teachers as they implement the components. The process will NOT work unless the building administrator is present and supportive throughout four to six months of the project.

If several schools are participating, it is helpful if each school has a group of about 4 teachers who will volunteer to participate, so they can collaborate, assist, and encourage one another. Also, nobody should be forced to participate in the project. Since it requires both a personal and professional commitment, it should be voluntary, not something forced upon teachers.

Management and accountability

The success of the experience is dependent on all partners providing ideas and suggestions and effort. Somebody, however, needs to be clearly in charge of managing the overall process. Maintaining communication, setting agendas, developing timetables, and coordinating activities are important considerations. Most frequently, this role has been managed by the professional development facilitator.

Within each school building, a <u>specific administrator</u> should clearly be charged with local logistics for rooms, times, materials, communication with the teachers and other administrators, and coordination of scheduling, substitute teachers, and so on. This person also will coordinate preparations for the sessions with the facilitator.

If external funding is to be used to facilitate the project, there are more responsibilities to consider. Funding for in-service projects in school districts often comes through Eisenhower (Title II) funding. Some Eisenhower support is awarded directly to school districts (LEAs, or Local Education Agencies). Other parts of it come through universities (IHEs, or Institutions of Higher Education). Since Eisenhower money is awarded on the basis of grants, it must be administered and reported on according to specified federal and state guidelines. The time to decide who will be responsible for signing purchase orders, keeping all parties informed about procedures, and organizing information for required reporting is when the project is initially designed.

In addition to fiscal accountability, **written reports** will be required. Determine at the onset what information needs to be accumulated to write these reports and who will take the leadership in preparing them. As with planning, all partners can provide ideas and effort for reporting. Data about the project can prove very helpful, especially when it comes to making presentations, whether it be to the school board or at professional meetings, or in professional publications. A well-designed **evaluation plan** will help meet all of these needs.

What sequence of events should occur?

Needs assessment

Before starting a project, and especially before writing a grant proposal to seek support for a project, it's a good idea to have some data that indicates the need and desire for the project. In this case, it may be a matter of identifying local requirements and expectations for teacher continuing education and curriculum review, as well as identifying potential sources of funding.

Information meeting

Ideas for the initial meeting have already been described. Much useful information can come out of this meeting. It is not just a presentation. It is a discussion of what the program can offer and how it fits into the local situation.

Focus, collaboration, agreement

Agree on commitments, dates, times, which classes to involve, and what grade levels to choose for modeling lessons. Set some specific goals and expectations within the context of the schools and the district. Look ahead to extension of the activities to other grades, classes, and schools. Anticipate relationships to district curriculum or pupil progression plan reviews, other projects, and grant opportunities.

How do we formalize the partnership?

Teachers make the commitment to participate and implement the strategies and philosophy in the classroom. Administrators agree to participate in the activities and support the teachers by providing release time (including paying substitute teachers if needed), buying materials, and providing travel money if needed. Paying for travel and the services of the professional development partner may also be necessary.

District regulations and procedures may vary on whether or not it is necessary or advisable to have actual signed agreements from teachers and administrators. There is, however, a need for a contractual commitment between the school district and the facilitator, in the form of a letter or contract stating the dates and nature of the in-service, and the amounts of the travel reimbursement and consulting fee to be paid.

When course credit is to be offered, the college or university may need advance commitment for scheduling of a course section, especially if the number of people enrolled is likely to be small. To encourage advance commitment of teachers, one project director asked each teacher to deposit a $25 check, to be held uncashed, and to be returned at the time of enrollment or to be forfeited if they didn't follow through. This method did help to keep the number of actual participants nearly equal to the estimated enrollment.

How do we seek funding support?

Sources of funding to support the in-service project also will vary from school to school. Most projects to-date have been supported substantially by federal funds, particularly from Eisenhower money and State Systemic Initiatives.

Districts and schools sometimes have other sources of staff development money. Numerous private foundations offer grants. Sometimes these are regional or discipline-specific. Often, they may offer priority to particular categories; e.g., girls/women, persons of a particular minority or ethnic heritage, persons with disabilities, relatives of employees of a sponsoring company, private school, inner city, rural. The "Seeking funds..." list that concludes this chapter provides ideas for a variety of kinds of support (Figure 13.1).

Colleges, universities, school districts, and state education departments all have personnel with expertise at locating potential funding sources and preparing grant proposals. It is their job to help people find ways to support worthy projects. They typically have much printed information to share. They also can assist in developing the budget part of a proposed project. In grant proposals it is usually necessary to indicate the amounts of cost-sharing among funding sources, institutional budgets, revenues, donations, etc.

Libraries often have information about federal grant programs. There also are two federal electronic databases--FEDIX and MOLLIS. They are free to use, but require some mastery. A particularly useful and user-friendly resource is SPIN-- Sponsored Projects Information Network--which offers more than 50,000 funding opportunities for grants, fellowships, scholarships, internships, and awards in a computerized database. SPIN and a similar service, IRIS, are subscription services. Many university grants offices subscribe to SPIN or IRIS, and your contact person there can help you obtain a search. Grant listings also are available at several sites on the Internet.

The bottom line is that there are many $$$ available to schools and teachers, especially in support of science and mathematics education. Sometimes-- especially at the district and state levels--there are relatively few proposals competing for available funds, because a relatively small number of eligible proposers know about and apply for them.

It's unrealistic to expect all of the funding needs for most projects to be supported by one source. Most grant applications require that you show some "matching" resources, both to emphasize your degree of commitment to the project and to expand what could otherwise be done with the grant money alone. Therefore, it's necessary to put together a budget and funding plan that coordinates support from all sources and effectively leverages various cash and non-cash resources to maximize the cash portion. To help express the value of non-cash matching items, a cash value must be estimated. Grant application guidelines often state what you may and may not count as matching support for purposes of the application.

If you decide to write a grant application proposal, think ahead. Check submission deadlines carefully and allow yourself long lead time. Also, find out the anticipated review and award schedule, as this will have an impact on your project planning. On federal grants, for instance, the review process may take 6 months, and it may take another 2-3 months before awarded funds are available. Plan ahead, and apply to more than one place at the same time to increase your chances for success. If you should be lucky enough to get support from more than one source, you will probably be allowed to negotiate some support from both places to meet matching commitments, fulfill your needs fully, or expand and extend the project.

The first grant proposal is the most difficult to write, but it is something with which you can find help. Your university professional development partner may be able to advise you well. Careful documentation and planning, creativity, and persistence will pay off. The second proposal is easier than the first, and so on. ☺

Some sources of funds & other tangible support

• **District and school budgets**
Basic operating budget, including some professional development money

• **Donations**
Money, equipment, books, supplies, magazine subscriptions, vehicles, water craft (for field work)

• **Fund-raisers**
Bake sales, car washes, auctions, book sales, garage sales, carnivals, benefit concerts, etc.

• **Ongoing money-making enterprises**
Bicycle and engine repair, storage sheds and birdhouses, restaurant, etc.

• **Booster clubs**
Often just for athletics and marching band. Some schools have booster clubs that are organized to support <u>all</u> student activities, including speech and drama, orchestra and choir, science fair, etc.

• **School support programs by local merchants**
Cash register rebates, donation to school with cable service sign-up, discount coupon books

• **Product rebates**
Campbell's soup and related labels, Community Coffee UPC patches, other

• **Use of volunteers**
Aides, field trips, special talks, expert advice, career information

continued on next page

- **Memorial gifts and bequests**

Scholarships, awards, books, physical facility items, collections, equipment

- **Government organizations, such as U. S. Army Corps of Engineers, state conservation, wildlife, and parks departments, U. S. Forest Service, U. S. Geological Survey, U. S. Park Service, NASA, U. S. Department of Energy, U. S. Bureau of Land Management, county USDA Cooperative Extension offices, others**

Grants, field trip planning and participation, loan of sampling and monitoring equipment, brochures and booklets, research publications, curriculum materials, teaching kits, posters, maps, slides and photographs, movies, videos, satellite images, classroom visitations, Internet sites, etc. This is a vast resource.

- **Grants and awards**

Local (school district and private home-town sources), **state** (requires some prospecting to identify), **federal** (_many_, including Eisenhower "flow-through" at district level), **private foundations** (such as Ford, Barbara Bush, Howard Hughes, Ohaus, Lindbergh), **corporations and corporate foundations** (such as Coca-Cola, power companies, Honda, CIBA-GEIGY, GTE, Amoco, US West), **organizations** (such as NSTA, American Association of University Women, Earthwatch, fraternal organizations), and **individuals** (may be local)

- **Other?**

Figure 13.1. Sources of funding for school-related projects. Adapted from workshop materials for "Getting $tarted on Grant$: Money to do What You Want to Do" (Saigo, 1994).

Author's note: If you have additional suggestions based on your own success, we would be pleased to add them to this list.

PART FOUR:
Appendices

Appendix I.

Interviewing

Sample student interview questions

Earth Science
seasons

Physical Science
magnetism
density and buoyancy

Mathematics
area and perimeter

Life Science
photosynthesis

EARTH SCIENCE

Seasons

1. What can you tell me about seasons?
2. How many are there? What are their names?
2. Do the seasons occur in a particular order?
    Follow-up question depends on student's response:
        What is the order? <u>or</u> Why is there no order?
3. Why are there different seasons?
4. How can you tell when it is winter?
5. Are there any other signs that suggest it is winter?
6. How can you tell when it is spring?
7. Are there any other signs that suggest it is spring?
8. How can you tell when it is summer?
9. Are there any other signs that suggest it is summer?
10. How can you tell when it is fall?
11. Are there any other signs that suggest it is fall?
12. What causes the seasons to change?

PHYSICAL SCIENCE

Magnetism

1. Have you seen magnets?
2. How do you know if something is a magnet?
3. What are some of the things that a magnet will pick up and some that it will not pick up?
4. Why will it pick up some things but not others?
5. What will you do to find out the strength of a magnet?
6. Can we make magnets?
7. What are some of the things through which a magnet can work?
8. What are some of the uses of magnets?

Density and Buoyancy

For #1:   Have the listed objects and a large, clear plastic aquarium that is 2/3 full of water to show the students.   They do not actually test their predictions during the interview; however, they commonly ask how much water will be used and it is important for them to see the concrete items when they are making their predictions.
For #2:   Have the described vials and a beaker of water.
For #3:   Have one piece of clay.   Form it into a ball, then into a boat.

1.   Which of these things (a grape, a watermelon, a bean, an ice cube, a block of ice, a sheet of aluminum foil, and a sheet of aluminum foil the same size that is crumpled into a loose ball) do you think will float and which ones do you think will sink in this much water?

> <u>Why do you think so?</u>
> (Then use follow-up questions based on what the student says.)

2.   If we place a closed vial filled with sand into a container of water, take it out, then place a closed vial of the same size filled with steel balls into the water, which of the following do you think will happen to the water level in the container?

> a.   Neither of the vials will displace any water.
> b.   Both vials will displace the same amount of water.
> c.   The vial with sand will displace more water than the vial with steel balls.
> d.   The vial with steel balls will displace more water than the vial with sand.

<u>Why do you think so?</u>
(Then use follow-up questions based on what the students say.)

3.   If we place a ball of clay in water, it sinks.
        If we make the ball of clay into a boat and put it into the water, it floats.

Which of the following things do you think will happen to the water level?
> a.   Neither the ball nor the boat will displace any water.
> b.   The ball will displace some water, but the boat will not.
> c.   The boat will displace water, but the ball will not.
> d.   Both will displace water, but the ball will displace more than the boat.
> e.   Both will displace water, but the boat will displace more than the ball.

<u>Why do you think so?</u>
(Then use follow-up questions based on what the student says.)

MATHEMATICS

Area and Perimeter
Use the same piece of string for making all of the shapes.

If we constructed the following shapes (a square, a rectangle, a circle) with this piece of string, which of the following things can you say about the shapes.

1.   The distance around will be
    a.   the same for all the shapes?
    b.   different, ranging from the largest to smallest as: [student gives order]?

      <u>Why do you think so?</u>
      (Then use follow-up questions based on what the student says.)

2.   The amount of coverage (amount of the desk/table top covered by the shapes) will be
    a.   the same?
    b.   different, ranging from the largest to smallest as: [student gives order]?

      <u>Why do you think so?</u>
      (Then use follow-up questions based on what the student says.)

LIFE SCIENCE

Photosynthesis

Use what you know about photosynthesis to answer these questions.
For #1: Try to have a plant in a large pot that simulates the original condition suggested for the student to look at. If you have "before" and "after" plants, be sure they are in the same size pots.

1.   The green plant weighed about 1 lb. without any soil on its roots before it was planted. Now, suppose we water the plant whenever the soil gets dry but do not add anything else to the pot. We take care of the plant this way for 5 years.

What do you think the plant will be like after 5 years?
Would it be bigger? smaller? the same?
Would it weigh more, less, or the same?

<u>Why do you think so?</u>
(Use follow-up questions based on what the student says.)

2.   If, after five years, we carefully removed the plant from the soil, then dried and weighed the soil, what do you think would have happened to the weight of the <u>soil</u> in the tub the plant grew in for 5 years?

a.   The soil lost weight.
b.   The weight of the soil stayed the same.
c.   The soil gained a lot of weight.

<u>Why do you think so?</u>
(Then use follow-up questions based on what the student says.)

Relationship between photosynthesis and respiration

3.   If a mouse were placed in a closed glass container, which of the following substances would you expect to <u>decrease</u> inside the container after the mouse had been living in there awhile: water, carbon dioxide, oxygen, nitrogen

<u>Why do you think so?</u>
(Then use follow-up questions based on what the student says.)

4.   If a green plant were placed in a closed glass container, which of the following substances would you expect to <u>increase</u> inside the container after the mouse had been living in there awhile: water, carbon dioxide, oxygen, nitrogen

<u>Why do you think so?</u>
(Then use follow-up questions based on what the student says.)

5.   Do you think the mouse could live in the closed container for very long?

      Why do you think so?
      (Use follow-up questions based on what the student says.)

6.   Do you think the plant could live in the closed container for very long?

      Why do you think so?
      (Use follow-up questions based on what the student says.)

7.   If a plant were placed in the container with the mouse, but protected so the mouse couldn't eat it, what do you predict would happen to the plant and the mouse?

      Why do you think so?
      (Use follow-up questions based on what the student says.)

(Adapted from Wandersee, 1986)

Appendix II.

Informed consent for participation in research

Sample consent form for adult subject

Sample consent statement for questionnaires and surveys

Special consent for minors

> **Sample parental consent form ***

> **Sample minor assent form ***

> *Must have BOTH for doing research involving a minor child.

Informed consent
for participation as a research subject

Sample consent form for adult subject

Federal regulations require that no one be required to participate as a research subject without their "informed consent." These regulations came about as the result of abuses in medical, military, and psychiatric research. The requirements of "informed consent" are that the subject be capable of understanding the terms of the consent and willing to voluntarily participate after procedures and potential risks and benefits are explained. If you have a question about what you are doing meets the federal definition of research, contact a grants officer at a nearby university.

Consent agreements that require signatures should never be physically part of any questionnaire, evaluation, or test document because that violates the required guarantee of confidentiality (or anonymity, if no one has a key to the individual subjects). If a signed consent is required, it must always be a separate document.

The following prototype statement contains the several required elements of informed consent. The language can be changed as long as the basic rights of the participant are stated. Begin the form with an explanation of the project (paragraph 1 below). Then state the required components as a part of the consent statement (paragraph 2 below).

> This study will investigate [state purpose of research]. With your help, we hope to [may state why their participation is important, potential benefits of the research]. If you have questions about the study, please contact [name, address, phone number of researcher].
>
> By signing this statement, I understand that the results of the study may be [insert how data will be used; e.g., for publication, used by committee or commission, used by the school board, etc.]. No data will be used in such a way that it reveals my identity. I understand that my participation is voluntary, and that I may ask questions or withdraw participation at any time with no penalties or loss of benefits.
>
> ......Signature and Date

Sample consent statement for questionnaires and surveys

The following is a statement that can be printed at the top of the questionnaire. If it is "anonymous" it does not require a signature. If it is a "confidential" survey and it is important to have a way to identify the respondent, you will need a separate, signed consent form. NEVER have the signed consent as a part of the written research instrument. Use a coding system on the individual questionnaires, surveys, or interviews and keep the key to the code secure.

> We would appreciate your participation in a study we are conducting about [......]. Your responses will be not be reported in a way that can identify you personally. By completing this questionnaire you are providing informed consent for the use of the data you provide.

Special consent for minors

You must have both parental consent and the agreement (assent) of the child for doing research involving a minor child.

Sample parental consent form

Be sure to explain the project to school authorities and get their permission to be in the school conducting the study. If parents have questions, they will surely call the principal. NOTE: If there is a likelihood of risk or discomfort, seek advice about protocol. The statement below assumes there is none. Also, the statement assumes you won't be sharing audiotapes or videotapes, which would permit identification of the subjects. Transcripts are OK as long as the subjects can't be identified, by name or by a combination of demographic information; e.g., being the only boy in the group, or the only 7th grader taking geometry, etc.

Begin the form with an explanation of the project (paragraph 1 below). Then state the required components as a part of the consent statement (paragraph 2 below).

We are conducting a study at [name of school] about [research topic], and would like your permission to include your child in the study. With the data we collect, we hope to [explain potential benefit of the research]. There will not be any risk or discomfort to your child during the study. If you have questions or would like additional information about the project please contact [name, address, phone number]. Thank you.

As parent or legal guardian, I [name of person] give permission for [name of minor] to participate in the study of [research topic] that is being conducted by [name of researcher]. I understand that the participation is voluntary, that there will be no penalty or loss of benefit to my child is he/she does not participate, and that I may ask questions or withdraw my permission at any time. Also, no data will be reported in a way that could identify my child.

Signature and Date

Sample minor assent form

I, [name of child], am aware that my parents have given permission for me to participate in a study concerning [topic of study] under the direction of [name of investigator] from [name of university or other organization]. I understand that my participation is voluntary and that I may ask questions about it or decide not to participate (withdraw my consent) at any time without penalty or loss of benefits to myself.

......Signature and Date

Appendix III.

Ideas for observation and reflection

Overview of observations and reflections: what to track

Sample journal entry combining observation and reflection

Questions to guide reflection regarding:

   Modeling

   Concept/topic

   Implementing the CCM

   Peer coaching

   Personal educational philosophy

Overview of observations and reflections: what to track

As we continue the process, please keep track of the following:

- Reflections about changes in your <u>expectations</u> of students.
  Use one concept as an example.

- Reflections about <u>experiences</u> you provide.
  Are you providing your students different experiences? Please give examples.

- Reflections on <u>assessment</u>.

<u>What to assess</u>
knowledge
conceptual development
skills
applications
attitudes
values
other

<u>Means to assess</u>
tests
interviews
parents
observing students
portfolios
projects
other

- Reflections about your own thinking about the <u>teaching/learning process</u>.

- What will <u>you tell</u> a parent that you are doing differently and why?

Ideas for journal entries
combining observation and reflection

Example A. Stating a specific plan/focus

Today I'm going to look for and think about:
1.   kinds of questions the students ask
2.   how I respond to them
3.   why they might be asking the kinds of questions they are asking

Example B. An efficient way to take notes and link them to reflections

Divide the page vertically, in a steno-pad sort of format. As the day goes along, jot down your observations (data) on the left. Later, you can go back and reflect on the observations to the right of them. Using this method, you preserve the spontaneity and accuracy of your observations and also relieve the burden of having to try to find time "sometime later" to remember it all and write it down.

Observations	Reflections
Every child volunteered a prediction	I think the practice we have done-- all the time taken last week is finally paying off.
4 of the groups worked well together, took turns in the group doing the hands-on part of the experimenting	The pendulums activity seems to have broken the ice because the kids wanted to and were able to test them over and over again. It was something all of them could do, the "test runs" were quick, and they were all involved to be sure it had been "done right" so they could believe what they were seeing.

Questions to guide reflections
following lessons that have been
modeled with students
(for teachers and administrators)

- What topic was addressed?

- What teaching strategy was modeled?

- How did the children react to the lesson?
  [Provide specific examples.]

- What were the strengths of the lesson?

- What were the weaknesses of the lesson?

- How was what you observed different from what you anticipated?

- What are your concerns about implementing the strategy in your own classroom/school?

Suggestions for reflections regarding the topic and/or concept to be taught
(for teachers)

Although this can be a hand-in assignment, it's best put in the journal.

Name:
Name of partners in the group:
Name of topic:

- Initially I thought the following about this topic:
  [include all aspects of the topic discussed]

- The things I learned from group and class discussion were:

- I changed or did not change my views and thinking with respect to the topic, because:

- What helped me to change my opinion/belief were:

- Some of the things that are/were difficult for me to change are:

- This is how I now understand the topic:

- The things that I learned about students' misconceptions on the topic, background information, and others are:

- These are some of the applications and outside-class examples of the topic:

- Some other questions and problems related to the topic that could be pursued are:

- Some of the things I still do not understand about the topic are:

- What I found valuable today that I can use with my students:

- Other thoughts:

Questions to guide reflections about myself and children while implementing conceptual change lessons
(for teacher)

A.   Planning Phase

   1.   What is the topic of the lesson?

   2.   What are the key points I want to address?

   3.   What do I expect of students?   [Include all domains.]

   4.   What are some of my students' misconceptions about this topic?

B.   Implementation Phase

   1.   Were the students able to make predictions and give explanations?

   2.   Were the students willing to make predictions?

   3.   Were the students willing to share ideas and listen to others?

   4.   Were the students willing to test their predictions?

   5.   Were the students willing to accept their observations whether or not they matched the predictions?

   6.   What changes did I observe regarding students' conceptual understanding related to the topic?

   7.   What changes did I observe regarding students' attitudes towards the lesson? [Consider students' confidence, pursuit of study, and reaction to being challenged.]

   8.   What changes did I observe regarding students' use of science process skills?

Questions to guide reflection
about peer coaching
(for teacher being coached)

- What did I want my colleague to look for?

- How did I feel about the lesson as I was being observed?

- How did the feedback from my coach help me improve this particular skill or strategy?

- What are some of the specific changes I would make if I were to teach the lesson again?

Questions to guide reflection
about peer coaching
(for teacher serving as coach)

- What did my colleague want me to look for?

- What were my feelings as I was observing the lesson?

- How did the feedback session help me in practicing the targeted strategy/strategies?

- What are some of the specific steps I need to take in order to improve my coaching skills?

Questions to help reexamine
personal educational philosophy
(for teachers and administrators)

- As a result of observing and talking with students, sharing with colleagues, implementing new strategies, and examining curricular materials, what changes have occurred?

- What now is my educational philosophy?

Appendix IV.

Sample lessons using the teaching for conceptual change model (CCM)

Physical Science
   pendulums*

Mathematics
   area and perimeter*

Integrated science and mathematics
   levers*

Earth science
   weather

Life science
   water movement in plants*

*Printed with permission of the authors and/or publishers, as indicated. See statement of copyright and use restrictions on next page.

Copyright and use restrictions on sample lessons

A. The CCM lessons on <u>Pendulums</u>, <u>Area and Perimeter</u>, and <u>Levers</u> are from:

> Stepans, Joseph. 1994. *Targeting Students' Science Misconceptions: Physical Science Activities Using the Conceptual Change Model.* Riverview, FL: Idea Factory, Inc.

This material is being reprinted with permission from the publisher and <u>may not be reproduced for classroom use without permission</u>. Please contact **Idea Factory, Inc.**, 10710 Dixon Drive, Riverview, FL 33569 (Telephone 800-331-6204), for additional information.

In addition to the activities, the chapters in *Targeting Students' Science Misconceptions* identify commonly held misconceptions and appropriate scientific explanations of the concepts in each topic.

These lessons are good examples of a fully developed lesson using the teaching for conceptual change model. Note that they

- clearly identify a topic,
- base the lesson on student activities and thinking,
- provide a series of experiences (activities) to help the student to confront and change existing beliefs,
- use a minimum of terminology, and
- actively engage the students in real investigations.

Furthermore, they are not complicated to set up and carry out. Based on your own grade level and experiences with your students, you can adapt them for older students.

It is interesting to note that *these lessons have been successfully taught "as is" with high school, college, and adult groups.* As described in the literature on misconceptions, adults typically have many of the same ideas children have about phenomena. We still cling to the views we formed as children unless something specific and compelling happens to cause us to undergo conceptual change. The reactions of adults to these lessons are often exciting.

B. The lesson on <u>Water Movement in Plants</u> is part of a work-in-progress on conceptual change lessons for life science by Barbara W. Saigo (© 1995 Saiwood Biology Resources). We welcome you to copy and use the lesson (with attribution of source) and forward your comments, criticisms, and suggestions to the address indicated on the title page of this book.

ACTIVITIES: Period of the pendulum

ACTIVITY I. Pendulums with the Same Length but Different Mass

1. Commit to an Outcome

Consider two equal-length pendulums, one with a wooden ball for a bob and one with a steel ball as a bob. (See Figure 1 below.) Clearly, the steel ball is more massive than the wooden ball.

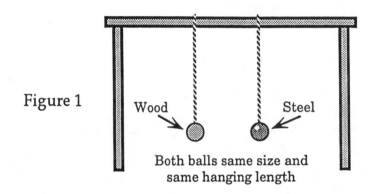

Figure 1

Wood   Steel

Both balls same size and
same hanging length

Predict what will happen if the two pendulums are set in motion at the same time. Which of the following is true for the <u>first few</u> swings (3-4):

    a. The wooden ball will swing faster (make more swings in a given time).
    b. The steel ball will swing faster (make more swings in a given time).
    c. They will have the same speed (make the same number of swings in a
       given time).

Make individual predictions and give explanations for the predictions.

2. Expose Beliefs

Share your predictions and explanations about the two pendulums. After listening to the predictions and the explanations of others in the group, do you want to make any changes in your thinking? Select a representative to present the predictions and explanations of the members of the group to the large group.

3. Confront Beliefs

Get the bobs and string and set up the pendulums to test your predictions. Rethink your explanations, if necessary. You may want to do this several times and have different people count periods (to 3-4) of the different pendulums. Remember we are looking to compare the *period* of the pendulums — which will go over and back faster?
What did you observe? How did your observations compare with your explanations?

4. Accommodate the Concept

Based on your observations and discussions, what statement can you make about the period of pendulums with equal string length and equal size bobs, but with one bob being much heavier than the other? Share your statement(s) with others in the group.

5. Extend the Concept

Can you think of other examples of equal size pendulums with different masses?

ACTIVITY II. Pendulums with the Same String Length but Different Shapes

1. Commit to an Outcome

Shown in the Figure 2 below are two pendulums, hung from strings of the same length.
Notice that the bobs have different shapes. One is a wooden sphere, as before, while the
other is a wooden cylinder.

Figure 2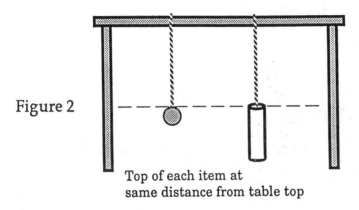

Top of each item at
same distance from table top

If the two are released at the same time, which of the following will happen? Give reasons
for your predictions.

      a. The sphere will swing faster.
      b. The cylinder will swing faster.
      c. The two will have the same speed.

2. Expose Beliefs

Share your predictions and explanations with others in your group about what will happen
to the two pendulums. Have a representative from your group share the predictions and
explanations with the members of the class.

3. Confront Beliefs

Get the sphere, cylinder, and string, and set up the two pendulums. As before, test your
predictions by counting 3-4 swings over and back and finding out which pendulum gets
there first. Again, you may want to do this several times.

What did you observe? Did your observations agree with your predictions?

4. Accommodate the Concept

Based on your discussions and observations of the sphere and the cylinder, what statement can you make about the period of pendulums with equal string length but different shapes?

Do you want to make any changes in your statement from Activity I based on what you have observed and discussed here?

5. Extend the Concept

What are some of the applications of what you have observed here? Can you think of examples related to these applications?

ACTIVITY III. Pendulums with the Same TOTAL Length

1. Commit to an Outcome

In Figure 3 below, there are two pendulums with equal total lengths; in other words, the distances from the points of suspension to the bottom of the bobs are the same. As before, the bob of one is a wooden sphere, and the bob of the other is a wooden cylinder.

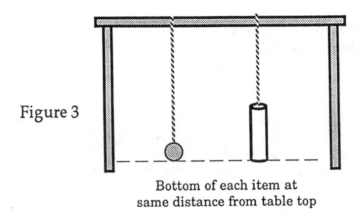

Figure 3

Bottom of each item at
same distance from table top

Again, predict which of the following will happen if the two are released at the same time and give reasons for your predictions.

      a. The sphere will swing faster.
      b. The cylinder will swing faster.
      c. The two will have the same speed.

2. Expose Beliefs

Share with others in your group your predictions and explanations as to what will happen if we swing two pendulums of equal total lengths.

3. Confront Beliefs

In your group, test your predictions about the periods of two pendulums with equal total lengths. What did you observe? Did your observations agree with your predictions? Do you want to make any changes in your thinking?

4. Accommodate the Concept

Based on your observations and discussions, write down a statement about the period of two pendulums which have different shapes but equal total length. Utilizing your observations and discussions from Activities I, II, and III, what statement can you make about factors which determine or do not determine the period of a pendulum?

5. Extend the Concept

What are some of the examples related to this with which you are familiar? Can you think of applications of this phenomenon?

ACTIVITY IV. Pendulums with the Same Length to the Center of Mass but Different Masses and Different Shapes

1. Commit to an Outcome

Consider Figure 4. There are two pendulums with different total lengths and different string lengths, but the distance from the top (the point of suspension) to the center of the masses is the same. Again, the bob of one is a wooden sphere, and the bob of the other is a wooden cylinder.

Figure 4

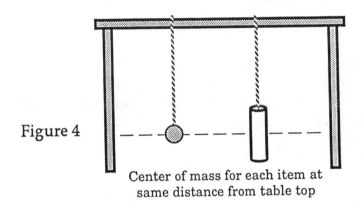

Center of mass for each item at
same distance from table top

Once again, predict which of the following will happen if the two are released at the same time and give reasons for your predictions.

      a. The sphere will swing faster.
      b. The cylinder will swing faster.
      c. The two will have the same speed.

2. Expose Beliefs

Share with others in your group your predictions and explanations about the period of these two pendulums where the distance from the point of suspension to the center of mass is the same.

3. Confront Beliefs

In your group, set up the two pendulums and test your ideas. You may repeat this several times. What did you observe? Did your observations agree with your predictions?

4. Accommodate the Concept

Incorporating your observations from all four situations, write a cumulative statement about what you think is the determining factor affecting the period of a pendulum. Share your statements with others in class.

5. Extend the Concept

Share examples and situations related to what you have learned here. What are some of the applications of what you have observed and discussed? What are some of the examples that you have encountered in your own lives or in other classes? Pendulums are common, and you may bring up any situation that is familiar to you.

6. Go Beyond

Think of other questions or problems related to pendulums that you may be interested in pursuing outside the class.

Some common misconceptions about the behavior of pendulums

- Mass ("weight") is the primary factor determining the period of a pendulum.

- Some students will believe that the pendulum with the lighter bob will move faster, while other students will believe the heavier one will move faster.

- The period of the pendulum is often confused with the speed of the pendulum. In fact, the speed depends on both the effective length and the angle of deflection.

- The string length alone is the important contributor that determines the period of the pendulum.

- What the pendulum is made of determines the period of the pendulum.

- The period of the pendulum is the same as how long it swings before it stops.

- The shape of the pendulum determines its speed.

- Some students cannot distinguish the effects of gravity, air resistance, and friction from factors that affect the period of the pendulum.

Area and Perimeter

ACTIVITY I: THE FENCE

1. Commit to an Outcome

Consider the following situation:

A farmer had a piece of fence. He used it to enclose part of pasture, as shown in Figure 1a below:

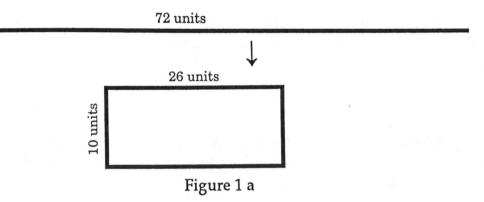

Figure 1 a

The following year, he took the same piece of fence and enclosed another part of the pasture, as in Figure 1 b:

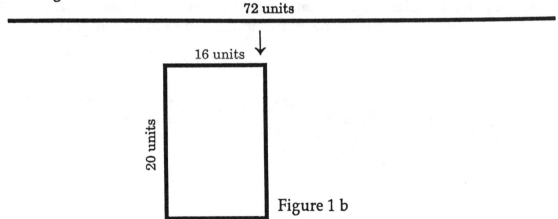

Figure 1 b

He was puzzled and asked himself these questions about the second year pasture:

    a.   Do I have more distance to walk around?
    b.   Do I have less distance to walk around ?
    c.   Do I have the same distance to walk around?

The farmer also wondered:

    d.   Do my cows have more grass to eat?
    e.   Do my cows have less grass to eat?
    f.   Do my cows have the same amount of grass to eat?

Select the options with which you agree and give reasons for your beliefs.

2. Expose Beliefs

Share your predictions and explanations in your small group. Based on your sharing with others in your group, do you want to make any changes in your ideas? Select a representative to present the predictions and the explanations of each member to the rest of the class.

3. Confront Beliefs

Get the necessary materials and test your predictions. What did you observe? Did your observations agree with your predictions? Any surprises? Do you want to make any changes in your thinking?

4. Accommodate the Concept

Based on what you have observed and the discussions in your group, what statement can you make about what happens to length and area when shape changes? Share your thoughts and statements with others.

5. Extend the Concept

If, from a given length, you made the shapes of a rectangle, square, and circle, how would the areas compare? What are some of the applications of this concept?

6. Go Beyond

Think of additional questions, problems, or projects you want to pursue related to the geometric concepts covered here.

Levers: Integrated science and mathematics

ACTIVITY: Exploring Levers

I. Situation One

1. Commit to an Outcome

Look at Figure 1 below. On the right, there are 2 weights located 3 spaces from the fulcrum, and on the left there are 3 weights located 2 spaces from the fulcrum.

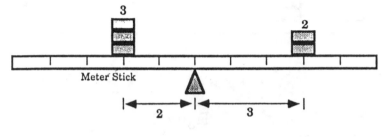

Figure 1

Predict which of the following will happen if you set up such a situation:

   a. The weights will balance.
   b. The lever will tip to the right.
   c. The lever will tip to the left.

Give explanations for your prediction.

2. Expose Beliefs

Share with others in your group your predictions and explanations as to what will happen
to the lever. Have a representative from your group present to the large group the predictions and explanations of the individual members.

3. Confront Beliefs

In your group, get the necessary materials, set up the situation in the figure, and test your
predictions. What did you observe? Did your observations agree with your predictions?
Do you want to make any changes in your thinking about levers at this point?

4. Accommodate the Concept

Based on your observations and discussions, what statement can you make about the
levers? Share your thinking with others.

II. Situation Two

1. Commit to an Outcome

Now examine Figure 2 below. On the right we have 2 weights, 4 spaces from the fulcrum, and on the left we have 3 weights, 3 spaces from the fulcrum.

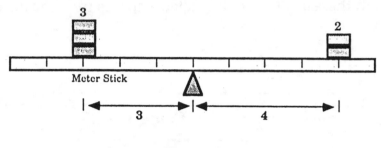

Figure 2

Predict which of the following will happen:

    a. The lever will balance.
    b. The lever will tip to the right.
    c. The lever will tip to the left.

Give an explanation for your prediction.

2. Expose Beliefs

Share with others in your group your predictions and an explanation about what will happen to the lever in this case. Have someone from your group share with the large group the predictions and reasons of your group's members.

3. Confront Beliefs

Use the weights and the lever as in the figure, and test your predictions. What did you observe? Did your observations agree with your prediction? What changes do you want to make in your thinking, if any?

4. Accommodate the Concept

Based on your observations and group discussions, what statement do you want to make about the levers? Share your statements with others.

III. Situation Three

1. Commit to an Outcome

Consider now the third case, as shown in Figure 3 below. On the right we have 2 weights, 3 spaces from the fulcrum, <u>and</u> 4 weights, 6 spaces from the fulcrum. On the left we have 7 weights, 5 spaces from the fulcrum.

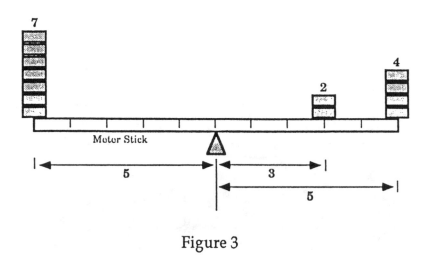

Figure 3

Predict which of the following will happen to the lever in this case:

    a. The lever will balance.
    b. The lever will tip to the right.
    c. The lever will tip to the left.

Give an explanation for your prediction.

2. Expose Beliefs

Share with others in your group your predictions and explanations about what will happen to this lever. Again, have a representative from your group present to the class the predictions and explanations of your group members.

3. Confront Beliefs

As before, test your predictions by constructing the lever pictured. Revise explanations, if necessary. What did you observe? Did your observations agree with your predictions? What changes do you want to make in your thinking?

4. Accommodate the Concept

Based on what you observed so far and your discussions, what statement can you make about what is involved in the behavior of levers? Has your explanation changed? Share your statements with others

IV. Situation Four

1. Commit to an Outcome

Look at the final figure (Figure 4). On the right we have 6 weights, 6 spaces from the fulcrum and 3 weights, 2 spaces from the fulcrum. On the left we have 3 weights, 10 spaces from the fulcrum and 4 weights, 3 spaces from the fulcrum.

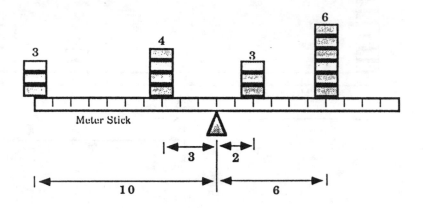

Figure 4

Predict which of the following may be true for the lever in the last figure:

      a. The lever will balance.
      b. The lever will tip to the right.
      c. The lever will tip to the left.

Provide an explanation for your prediction.

2. Expose Beliefs

Share with others in your group your prediction and explanation about what will happen to the last lever. One last time, have a representative report the views of your group members to the large group.

3. Confront Beliefs

In your group, again use the necessary weights to set up the lever as shown, and test your predictions. What did you observe? Were there any surprises? Do you want to make any changes in your thinking about what is involved in levers?

4. Accommodate the Concept

Based on your observations in all four situations, what general statement can you make about the rule of levers?

Check your general statement by applying it to each of the four situations. If it does not work, try to come up with a different rule and check that one with the weights and lever.

5. Extend the Concept

Test your rule by setting up other situations that you invent. What are some examples of this concept with which you are familiar?

Explain teeter-totters, the process of opening a door, and the operation of jack handles in terms of what you have learned here.

6. Go Beyond

What additional questions, problems, or projects would you like to pursue related to levers and their applications?

Weather: Clouds

This conceptual change lesson is designed for young children.

1. Commit to an Outcome

This afternoon as we continue our study of weather, we will go outside and look at the sky. We will be observing the clouds, if there are any. Right now, I want you to predict what we will see. You might want to think about what the weather was like when you came to school this morning. If you like, you may take a minute to go to the window and look at the sky as it is now. If, in four hours, we go outside and look up at the sky, what do you think we will see?

2. Expose Beliefs

Draw a picture or write a few words describing what you think we will see. Share that with the others in your group. It's okay if you have different ideas. [allow time]

Do the others have similar predictions? Are there big differences in the predictions? Have a spokesperson for your group share all of the ideas from your group with the entire class.

3. Confront Beliefs

Now that several hours have passed since we made our predictions, let's go outside and look at the sky. Take a pad of paper so that you may write down some observations or draw pictures of what you see.

NOTE: Be sure the students know what "observe" and "observation" mean.

4. Accommodate the Concept

Let's sit down outside and talk about what we have seen/observed. Are there any clouds in the sky? How much of the sky is taken up by the clouds? What color are the clouds? Do the clouds appear to touch one another? How do these observations compare to your predictions?

5. Extend the Concept

Have you ever noticed the sky before today when the clouds looked very much like this? What was the weather like at that time? Was it rainy? Was it sunny? Was it hot, warm, or cold? Did the weather change very quickly, or not? What do you think the weather will be like for the next 24 hours [or- between now and this time tomorrow]?

Compare these new predictions with the weather conditions the next day.

6. Go Beyond

Keep a record of the sky observations for the next week or so.

- Record the color and shapes of the clouds.
- In addition, record the amount of the sky that is covered by clouds: mostly clouds, half or partly cloudy, only a few clouds, or no clouds.

Also keep a record of other weather conditions, such as:

- General conditions: rainy, sunny, gray day (no sunshine but no rain either)
- Temperature: hot, warm, cool, or cold
- Air movement: very windy, some wind, occasional breeze, no wind

After a period of time--perhaps 2-3 weeks--compare the observations related to clouds with observations of the other conditions.

- Are certain kinds of clouds visible when it is sunny and warm?
- Are there certain kinds of clouds visible when it turns cold or windy?
- Are their certain kinds of clouds visible before it rains? or snows?
- Do those clouds look the same as the clouds which actually rain?

Water movement in plants

The Activities in this set are <u>progressive</u>, moving

- from developing the concept that water is in plants and plays a role in supporting their tissues

- to development of the concept that it moves through plants through special pathways (not just in general, like through a sponge),

- to the relationship that loss of water from the leafy parts plays a role in this movement.

They are progressive not only in developing a series of related <u>concepts</u>, but also in terms of guiding the students into increasingly sophisticated <u>experimental design</u>.

The level of terminology that can accompany the development of this series of concepts can vary with the level of the learner. However, <u>the concepts themselves should drive the activities, not the terminology</u>. The learners should be able "invent" the concepts of turgidity, veins and/or vessels, water transport, and transpiration THEN find the names for these ideas, not the other way around, or the lab work is reduced to the traditional confirmation format.

The interview questions can be used for middle elementary to high school students. The activities can be modified as appropriate to the age and experience of the students.

This lesson is part of a work-in-progress on conceptual change lessons for life science (B. W. Saigo, ©SBR, 1995). <u>Comments are welcomed</u> as per page A-24 request.

Pre-interviews

Before beginning this topic, interview a few students to determine a probable starting level of understanding and misconceptions regarding this topic. Some possible questions are:

- What can you tell me about water in plants?

- Where do you think the water in plants comes from?

- What happens to water after it gets into a plant?
  Where does it go? What does it do?

- How does the water move inside the plant?

- How does water get out of the plant?

- What happens to a plant if too much water gets out of it?

- Do you think there is a relationship between how fast water gets out of a plant and how fast it gets into it? Tell me what you mean.

<u>Do not introduce scientific terminology during the interview</u>. Allow the students to make their explanations in their own words, so you can gauge their understanding of the <u>concepts</u>, independent of vocabulary you might use during instruction.

For the record: <u>A celery stalk is not a stem</u>. That is a misconception. The part we usually eat is a leafstalk, or petiole. The leaf blade at the tip is divided into leaflets. The base of the stalk attaches to the stem of the celery plant. The stem is the celery "heart," which is basically cone-shaped. When you examine the celery stem, you can see that all the leafstalks grow from it in a spiral arrangement in which new leaves continue to be produced from the center, where the apical meristem is located. If you bisect the celery stem you can relate its structure to other dicot stems; but that is material for another topic, not this one.

TOPIC: Water movement in plants

Activity I: Water and support

As a <u>first</u> activity to set up the <u>concept of water in plants</u>:
Provide each group a wilted stalk and crisp stalk of celery, both with leafy tops, about the same size. Explain they can handle the stalks to determine their condition, but be careful not to break anything off.

1. Commit to an Outcome

How can you explain the difference in stiffness between the two stalks of celery? Do you think you could make the wilted one crisp and the crisp one wilted? What would you have to do and how long do you think it would take, in each case? Design ways to test your ideas.

2. Expose Beliefs

Working in your small group, write down possible explanations for the differences in the condition of the two stalks. Share your predictions and how you will test your ideas. Then, have a spokesperson share your group's ideas with the large group.

3. Confront Beliefs

Get beakers or other containers with which to conduct your experiment. Cut about
1 cm off the bottom end of each stalk, or have your teacher do it. Weigh both stalks of celery after you have trimmed them. Record the starting data (for example: weight, water level, descriptions) and time, then begin your experiment. Make observations every 10 minutes and record the time and new data. Record the data from your group on a large chart along with data from all of the groups.

<u>Teacher</u>: One way students may perform these tests is to stand the wilted stalk in a container with water and stand the crisp stalk in a similar posture in an empty container. Weighing before and after and recording water level before and after provides information for inferring water went out of the container into the celery. Both set-ups should be kept side-by-side, under identical environmental conditions.

4. Accommodate the Concept

Share your conclusions about what causes the celery to be firm (turgid). How does what you observed compare to your original ideas? Make any changes to your original explanation that are needed based on the activity.

Activity II: Path of water movement

1. Commit to an Outcome

In Activity I, you saw that water can move into plants. How do you think the water moved in and throughout the tissues of the plant after it got inside? Can you think of an analogy from your experience? Suppose you were to place a stalk of celery in a container of colored water. Predict where you think the dyed water would appear inside the plant. You may want to make a drawing to show the mental model of where the colored water will go. Explain your prediction.

2. Expose Beliefs

Share your predictions and ideas with the other members of your small group. Have a member of your group share all predictions and ideas with the large group.

3. Confront Beliefs

Perform your experiment and record your observations. You may trim the bottom end of the celery stalk to see where the dye is located after 5-10 minutes. Make a drawing of what you see.

Put the celery stalk back in the liquid and observe it again in 20 minutes. Is the dye apparent in the leafy portion? Is it distributed uniformly? If not, how is it distributed?

Return the celery to the dye and leave it overnight, then look at it again during your next class session. After you have recorded your observations, you may pull out some of the celery "strings." See if you can pull a single "string" the full length of the stalk. Examine the string and the other tissues inside the celery stalk. Describe and explain what you see.

4. Accommodate the Concept

Based on all of your observations, what can you conclude about the pathway of water movement in the celery plant? Compare your observations to your initial predictions and modify your explanation if needed.

5. Extend the Concept

What are some other examples and applications of what you have just observed?

6. Go Beyond

What might you do to explore further the structure and function of the celery "strings?" Why do you think they are called vascular bundles? What do you think is the shape of the cells of which they might be composed?

Teacher: Activity II can be repeated to yield quantitative data. It will involve cutting (paring knife, scalpel or razor blade) the ends of the celery at intervals, so the age, maturity, dexterity, and safety of students needs to be considered. Take the celery out of the dyed water at 10 minute intervals. Cut petiole segments against a grid to measure how far up the dye has moved. After a series of readings, calculate the rate of movement of the water.

Activity III. Leaves and water uptake

1. Commit to an Outcome

Do you think the leafy part of the celery stalk has anything to do with movement of water in the plant? Do you think the water uptake of a celery stalk with the leafy tip attached and one with the leafy tip cut off would be the same? Predict what you think would happen if you ran such an experiment. Explain the reasons for your prediction.

2. Expose your Beliefs

Share your predictions in your small group. Have someone write down all of the predictions and explanations. Then, share your group's predictions with the whole class.

3. Confront your Beliefs

In your small group, set up an experiment to test your prediction (hypothesis). Identify the possible variables in your experiment, and try to control all of them so the only difference between the two celery stalks and what happens to them is the presence or absence of the leafy tip. Decide how you will measure the amount of water that is taken into the celery to provide the data that is most relevant to your prediction. (Hint: The thinner the container in which you put the celery, the easier it will be to see changes.) Decide how often you will take measurements and how you will record your data.

4. Accommodate the Concept

Prepare a graph from your results to help you analyze them. Share your results with those of other groups. Based on your results and the results of the other groups, what statements can you make about the effect of leaves on water movement into a plant? Do your results support your initial belief or have they caused you to modify it?

5. Extend the Concept

Can you think of any applications of what you have just learned? What are some examples of what happens to plants in your yard, garden, or home that might be related to this idea?

6. Go Beyond

What are some other things you might do to explore regarding leaves and the movement of water into a plant?

Activity IV. Water loss

Teacher: This is a culminating activity. It is more sophisticated and takes longer than the others. The extra time is due largely to the mental processing and synthesis required of the students. Students will require more time in their initial discussion of predictions, in planning their experimental design, and in synthesizing their concepts during steps 4 and 5.

1. Commit to an Outcome

In Activity I, you caused one of the celery stalks to become wilted from water loss. Where do you think the water went? How did it leave the plant?

Think about the following experiment:

> **A.** You have 3 stalks of celery, as similar as possible in size, shape, maturity (distance from center or "core" of the celery bunch), and number of leaflets at the tip.

> **B.** This is what you do to them:

>> 1.   Cut the leaflets off one stalk.

>> 2.   Put a plastic sandwich bag over the leafy part of the second stalk and seal it closed with tape.

>> 3.   Leave the leafy part of the third stalk uncovered.

>> 4.   Trim the bottom of all three stalks so they are the same length between the bottom and the crease where the leaflet bases branch from the main stalk.

5.   Stand the stalks of celery in three separate but identical containers of colored water (red food coloring), each with the same water level or volume.

C. Observe and collect data at intervals to see if any changes have occurred.

Predict what you would expect to happen to each of the set-ups in terms of:

- water level (volume) in the containers
- dye movement
- content of the sandwich bag
- how soon changes might occur
- any other factors you might like to check

Explain your predictions.

2. Expose Beliefs

Share your predictions and explanations with the other members of your group. Based on the discussion in your group, make any modifications you would like to make in your own predictions. Have someone write down all of the predictions, then share them with the larger group.

3. Confront Beliefs

Set up the experiments. Decide how you will collect the data needed to test your predictions. Collect and record data during the class period. Leave the experiments set up overnight. The next day at the beginning of the class period, make your final observations.

Prepare a table or graph of your data and analyze it. What are the important differences between the three stalks of celery in regard to your predictions?

4. Accommodate the concept

What conclusions can you make about how water moves out of a plant and where it goes? What factor or factors seem to affect the rate at which water is taken up and moved through the celery stalk? Does the data support your initial predictions and explanations? How have your observations changed your initial ideas?

Putting together all of the ideas you have explored in Activities I-IV, explain what your experiences have revealed to you about:

- The role of water in supporting the celery stalk.

- How water moves inside the celery stalk.

- How water leaves the celery stalk.

- The relationship of evaporation to water movement through the celery stalk.

Do you think any of your explanations can be generalized to all plants? Can any of them be generalized only to plants having certain characteristics? Explain.

5. Extend the Concept

What are some of the applications and other examples of these concepts you can think of? Think about days when the plants in your yard may have wilted. What were the environmental conditions on these days? What was the air temperature? Was the air dry or humid? Was it windy or still? What predictions might you make about the effect of each of these conditions on the loss of water from plant surfaces and what experiments would you design to test your predictions?

6. Go Beyond

What other questions or ideas for experiments can you think of regarding how water moves through plants? Are there related ideas or comparisons you could explore with other plants, such as a carnation flower? carrot? onion? cabbage leaf? cactus? pine or fir branch? grass leaves?

Appendix V.

Peer coaching forms

Use of conceptual change model (generic)

Use of JIGSAW strategy

during relative density lesson

Use of one component of the CCM linked to a specific topic

student predictions during erosion lesson --
"become aware" and "share" parts of CCM

Use of questioning

Peer Coaching Feedback Form
What is being observed:
Implementation of the Conceptual Change Model

Criteria	Observed (Example)	Not Observed
The teacher causes students to become aware of their own preconceptions about a concept by making personal predictions.		
The teacher allows students to expose their beliefs by sharing predictions and explanations in small groups and then with the entire class.		
The teacher allows the students to test their predictions and ideas by making observations of the phenomena. The students discuss the results of their test, as they work in small groups.		
The teacher creates an atmosphere where students can work toward resolving conflicts between their predictions and their laboratory observations by class discussion.		
The teacher provides opportunities for students to extend the concepts and try to make connections between what they have learned in the classroom and other situations.		
The teacher encourages the students to pursue additional questions and problems of their choice related to the concept.		

Peer Coaching Feedback Form
What is being observed: Use of JIGSAW strategy

Criteria	Observed	Did Not Observe	Example
The teacher places students in groups.			
The teacher assigns each group member to either white, green, brown, or yellow group.			
The teacher gives the group the task, which is to predict what will happen to a corn kernel when placed in a graduated cylinder containing various liquids.			
The teacher instructs members on their individual accountabilities. (Each member must learn the information about each liquid and explain to the rest of the group.) (Each member should be prepared to give the report to the rest of the class.)			
The teacher instructs the group to predict the order in which the liquids must be added to form layers and test their predictions.			
The teacher sets expectation that each member of the group should be prepared to predict what will happen to a solid when placed in the container with given liquids.			
The teacher instructs each individual to contribute to the solution of the task.			
The teacher sets criteria for success as the groups make correct predictions and test their predictions.			

Peer Coaching Example: Erosion
1. Scenario 2. Peer coaching form 3. Feedback

1.   Scenario

Teacher:

Since this is the first time I've taught a lesson using the Conceptual Change Model, I am going to focus on the first two steps of the model where students (1) become aware of their own beliefs and (2) share their beliefs. The unit is on erosion, but today's lesson focuses on the movement of water from a higher elevation to a lower elevation.

Specifically, I would like you to observe whether or not I elicit predictions from all children and if I avoid changing their predictions by using adult explanations.

2.   Peer coaching form generated by the teacher and the coach--see next page. The coach would observe the colleague teaching the class, attending specifically to the agreed-upon parameters, recording observations on this form. After the lesson, the coach would provide feedback.

3.   Feedback

Coach:

Your question, "What will happen to the water if we pour it slowly on the top of the pile of sand?" did stimulate students to predict outcomes.

In all of the small groups except one, the children shared their predictions. One child, in the group near the pencil sharpener, said nothing while the other three children talked a lot. In most groups the children took turns stating their predictions.

As children reported the predictions by group, you wrote the exact words in the key phrases of their explanations on the board. Examples:
1.   water will go away
2.   water will run down
3.   water will make a lake
4.   water will soak into dirt stuff
5.   dirt will get wet
6.   water and dirt stuff will go down
7.   water will gush down

2.   Peer Coaching Form--Erosion lesson (generated by the teacher and the coach)

Things to Observe	Evidence
1. Is the opening question stated in such a way that the children are encouraged to make predictions?	
2. Do all children make predictions?	
3. Does the teacher use the children's words rather t than adult vocabulary when recording predictions?	

Peer Coaching Example: Questioning
1. Scenario 2. Peer coaching form 3. Feedback

1.   Scenario
Teacher:

I want to work on my questioning technique. The lesson is on the relationship between color and heat absorption.

I want you to look for my use of open-ended questions that encourage divergent thinking rather than convergent thinking. I want to involve as many students as possible in the discussion. And, one more thing, I want you to look for wait time. I'm not really sure how much time I allow between questions and calling for a first response.

2.   Peer coaching form generated by the teacher and the coach. See next page.
The coach would observe the colleague teaching the class, attending specifically to the agreed-upon parameters, recording observations on this form. After the lesson, the coach would provide feedback.

3.   Feedback.
Coach:

These are examples of <u>open-ended questions</u> you used:
     Q1   What do you think will happen if we put a cork in the top of the
green   bottle and place it in the sunlight?
     Q2   Why did it do that?
     Q3   What observations help you explain that?
     Q5   How could we design an investigation to help answer the question?
     Q6   Do you think there will be a difference in the temperature inside
the   different colored bottles?

These are examples of <u>closed questions</u> you used:
     Q4   How do we measure temperature?
     Q7   Are there any patterns in the data collected?

Every student was <u>involved in the discussion</u> whether it was with a small group or with the entire class.

The amount of <u>time following each question</u> varied from 2 seconds to 65 seconds.
     Q1   25 seconds   Q5   65 seconds
     Q2   35 seconds   Q6   45 seconds
     Q3   15 seconds   Q7   50 seconds
     Q4   2 seconds

2.   Peer coaching form--questioning example.

Things to Observe	Evidence
1. How many divergent or open-ended questions were asked?	
3. How many students participated in the class discussion?	
4. How many minutes did the teacher wait before eliciting the first response?	

Appendix VI.

Lesson and assessment planning

Suggestions for planning learning materials

A continuing process of improving teaching and learning

Sample expectations

A list of possible student expectations for the pendulums topic.
* Note copyright statement below.

Aligning curriculum, instruction, and assessment

A matrix to go with the pendulums topic that matches <u>student</u> activities and expectations to appropriate <u>assessment</u> strategies and expectations
* Note copyright statement below.

Some other ways to assess

Including examples of assessment criteria

Suggestions for planning learning materials

Topic (concept):

Grade level intended:

Summary of the concept:

Students' preconceptions/possible misconceptions related to the concept (based on interviews):

Your expectations of students (include all domains) as the class continues with the concept:

The experiences you are planning to provide the students to help them meet the expectations. Include the conceptual change strategy.

Your assessment plan:

    How are you planning to assess

    1.   whether students have met the expectations (gear to all domains)

    2.   appropriateness of expectations

Plans for revision based on assessment results:

A continuing process of improving teaching and learning

- Initiate the concept(s)
- Decide what we should expect from the learner.
- Determine what the learner already knows and thinks about the concept(s)
- Review how the concept is usually presented
- Reflect on difficulties with teaching and learning the concept.
- Share observations
- Experience teaching strategies
- Develop own learning materials
- Implement the materials
- Collect data on students' learning, conceptual change, and attitudes
- Keep track of changes in own philosophy and perception
- Share observations and experiences
- Repeat the process
- In light of what is learned, evaluate curriculum, instruction, and assessment
- Align curriculum, instruction, and assessment

Sample Expectations

AN EXAMPLE OF APPROPRIATE ASSESSMENT

We will use the topic of pendulums to illustrate an example of appropriate assessment strategies. Let us begin by clarifying our expectations of students as we go through the topic. What is it that we hope to accomplish? Some examples of appropriate expectations are listed below. Feel free to add to the list or otherwise adapt it to your own student expectations.

Expectations

1. Students will become aware of their beliefs related to the behavior of the pendulum.

2. Students will become aware of the views of others.

3. Students will be willing to confront their ideas.

4. Students will be willing to revise their ideas.

5. Students will extend the concept of the pendulum as demonstrated in classroom activities to other situations and applications.

6. Students will go beyond with their knowledge of the pendulum outside of class, continuing to test their understanding.

7. Students will collaborate with others.

8. Students will show respect for the opinions of others.

9. Students will collect necessary data.

10. Students will identify and control appropriate variables.

11. Students will establish relationships among variables.

12. Students will initiate ideas.

13. Students will show persistence.

14. Students will communicate data and analysis of data using various modes (speaking, writing, tables, graphs, drawings).

15. Students will exhibit confidence in pursuing questions about the pendulum.

16. Students will appreciate the importance of the study of the pendulum.

17. Students will enjoy being challenged.

18. Students will develop a conceptual understanding of the pendulum.

Aligning curriculum, instruction, and assessment

Situation Presented	Assessment Strategy Used	Expectation Targeted	Expectations Met	Expectations Developing	Expectations Not Met
*Suppose that you have two pendulums, equal in size, and one has wooden bob while the other one has a metal bob. Predict what would happen if you pulled both of them toward you and released them at the same time. Provide reasons for your predictions.	*Observations *Interview (individual or group) *Performance	#1	*willing to make prediction *able to predict *willing to write down reason *able to come up with sound explanation	*attempts to predict *makes effort at reasoning *attempts explanation	*resists making prediction *makes no predictions *resists writing reason *unable to explain
*Share your predictions and explanations with others in your group and then with the class	*Observation *Performance	#2	*willing to share *able to explain *able to draw on experience	*makes attempt *tries to explain	*resists sharing *unable to explain *unable to draw on experience
*Test your ideas and make revisions if necessary	*Observation *Performance	#s 4,7,8,9	*willing to test *able to set up tests *makes accurate observations *collaborates with others *incorporates others' ideas *willing to revise ideas *shows persistence	*makes an attempt	*resists testing *unable to set up tests *unable to make accurate observations *does not collaborate *unwilling to revise ideas, *gives up easily
*Make a statement explaining your understanding of the behavior of the pendulums	*Obsevations *Pencil and paper	#4	*willing to write down ideas *able to synthesize information		*not willing to write down ideas *not able to synthesize information

Situation Presented	Assessment Strategy Used	Expectations Targeted	Expectations Met	Expectations Developing	Expectations Not Met
*Suppose you have two pendulums, one with a wooden ball for a bob and the other with a wooden cylinder. They are hanging from strings of equal length. Predict what would happen if you pulled them toward you and released them at the same time, comparing the time for 3-4 swings. Give explanations for your predictions	*Interview *Performance *Pencil and paper *Observation	#s 1, 2, 3, 4, 7, 8	*collaborates with others *shows respect for others' opinions	*attempts to collaborate *attempts to respect opinions	*resists collaboration *shows lack of respect for others' opinions
*Given a wood ball and a wooden cylinder hanging from strings such that the total lengths are equal, make predictions about the time taken for 3-5 swings. Give reasons for your predictions.		# 14	*willing to synthesize data from previous situation *willing to communicate concisely the behavior of the pendulum		*unwilling to synthesize previous data *unwilling to communicate behavior of pendulum

Situation Presented	Assessment Strategy Used	Expectations Targeted	Expectations Met	Expectations Developing	Expectations Not Met
*Using situations 1-4, make a statement about what factors affect the period of a pendulum.	*Pencil and paper *Observation *Interview	#s 11,14	*is able to establish relationships among variables *is willing to communicate data and analysis of data using tables, graphs	*attempts to establish relationships *makes an attempt to communicate data	*unable to establish relationships *resists communication and analysis
*How would you extend the topic of pendulums to other situations?	*Interview *Pencil and paper	#s 12, 13, 15	*initiates ideas *shows persistence *shows confidence	*attempts to initiate ideas *shows effort	*does not initiate ideas *gives up easily *lacks confidence
*Why do we study the pendulum?	*Interview *Pencil and paper	# 16	*volunteers to bring ideas about pendulums from home	*thinks about pendulum out of class *does not attempt to bring ideas to class	
*What questions, problems, or projects would you like to pursue about pendulums?	*Interview *Pencil and paper *Observation	#s 17, 18	*brings ideas for extension *exhibits a conceptual understanding	*shows interest in extending ideas *show some understanding	*does not show interest *does not show understanding
*Set up a plan to pursue your question, problem, or project. Present the results to the class.	*Performance *Observation	#s 5, 6	*is able to set up appropriate situations *shows interest *enjoys challenges	*makes attempt to set up situations *indifferent to challenges	*is not interested *resists challenges

Some other ways to assess
Including examples of assessment criteria

1.   Using science processes and laboratory activities

The student ...
*   states the problem
*   describes the experimental setup
*   uses lab safety
*   describes all variables
*   uses appropriate techniques to collect data
*   analyzes data appropriately
*   presents results clearly
*   clearly states how the results were reached
*   poses additional questions to pursue

2.   Cooperative learning

The student ...
*   cooperates with others
*   stays on task
*   respects the opinions of others
*   communicates with others
*   listens to others
*   contributes to the knowledge of the group members
*   revises opinions based on the ideas presented by other members
*   shows interest in the activity

3.   Working on a project

The student ...

*   identifies variables
*   conducts activities
*   sets up experiments
*   collects appropriate data
*   graphs
*   analyzes data
*   reports results
*   prepares a report

4.   Concept mapping

The student ...
*   constructs an initial map on the concept
*   discusses own map with others
*   consults written and other materials on the concept
*   takes part in activities relating to the concept
*   makes changes in the map based on what was learned from discussion
         with others, using the written materials, and engaging in the
         activities
*   explains the changes made in the map
*   identifies appropriate questions to pursue

5.   Student reflections

The student writes about the experience, including such reflections as ...
*   Initially this is what I thought may happen.
*   These are some of the things we did in our group.
*   These are some of the things I learned from others in the group and class
         activities.
*   This is how I now understand [the concept].
*   These are some of the questions I want to pursue.
*   These are some of the things I don't understand.
*   These are the steps I am planning to take to come to a better
         understanding.

6.   Jigsaw

A.   The expectations are related to the diagram representing the interactions
of the digestive, respiratory, and circulatory systems:

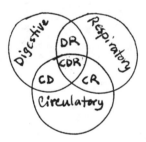

*   In each of the intersecting areas DR, CD, and CR identify the relationship of the two systems.

*   In the area CDR describe the interdependence of the three systems.

B.   Break into three groups. All members of each group will develop
expertise in one of the three systems--circulatory system, digestive system, and
respiratory system.

C.   After this is done, break into three new groups. Each group will have a
representative from the initial groups, so expertise on all three systems is
represented. The task is for each member of the new group to:
     1.   Learn about the relationship between any two systems
     2.   Be knowledgeable about the interdependence of the three systems

Appendix VII.

Ideas for student reflections

Sample topics for reflection

Portfolios

Suggestions for reflections regarding the topic and/or concept to be taught
(for students)

Name:

Name of partners in the group:

Name of topic:

- Initially I thought the following about this topic:
  [include all aspects of the topic discussed]
- The things I learned from group and class discussion were:
- I changed or did not change my views and thinking with respect to the topic, because:
- What helped me to change my opinion/belief were:
- Some of the things that are/were difficult for me to change are:
- This is how I now understand the topic:
- These are some of the applications and outside-class examples of the topic:
- Some other questions and problems related to the topic that could be pursued are:
- Some of the things I still do not understand about the topic are:
- Other thoughts:

Portfolios

Students may include in their portfolios samples of writing, drawings, lab reports, research ideas or results, photographs, graphs and charts, models, and other materials that reflect their understanding of the concepts and their ability to apply and explain them. Some portfolio contents may be artistic, such as poetry or graphics. Some may be humorous or whimsical.

The most significant point of using student portfolios is that the contents are selected by the student, who is able to explain the rationale for including each item. Student portfolios are distinct from folders of student work that have been assembled by the teacher.

Appendix VIII.

Evaluation questions

Questions to help improve implementation of the WyTRIAD

Interview questions, Kleinsasser and Miller, 1994

Interview questions, Stepans and Saigo, 1993

Questions to help improve the implementation of the WyTRIAD
(for professional development facilitator)

Which components are effective in creating positive change in the teaching/learning process?

Were the times suggested for teach session appropriate? If not, what changes need to be made?

How well did the partners fulfill their roles? What changes would improve the effectiveness of the partnership?

To what degree were the teachers able to implement targeted strategies and idea?

What were the barriers which inhibited the process and what can be done about them?

How should the process be evaluated?

What were the strengths of the project?

What were the weaknesses?

What were some of the unexpected and important things that happened during the project?

What are some of the steps that need to be taken to continue, disseminate, and expand the project?

Interview questions used for WyTRIAD evaluation
Kleinsasser and Miller, 1994

METHOD: Individual telephone interviews of teachers

1.   How has this project affected your science and mathematics planning and teaching? For example, what <u>specific changes</u> have you introduced into your instruction as a result of this project?

2.   What effects has <u>learning to interview</u> had on your teaching?

3.   Has this project (the interviewing, the focus on conceptual change, the role of the administrator, the peer support) affected your teaching in subjects other than science and mathematics? Please explain.

4.   In what ways has your participation in the TRIAD Project affected your relationships with other teachers and administrators?

5.   Please respond to the questions below from the perspective of being a teacher in a school or district that is considering implementation of the TRIAD Project.

     a.   What should I expect to learn?
     b.   What must I be prepared to do in order to maximize the goals of the project?
     c.   For the project to succeed, what factors must be in place in the classroom, school building, and district?
     d.   What information do learners need?
     e.   What information do parents need?

Focus Questions used during interviews of WyTRIAD teachers and administrators
Stepans and Saigo, 1993

METHOD: Group interviews of teachers by school,
plus group and individual interviews of administrators.

A. QUESTIONS FOR THE TEACHERS
(Does not include all follow-up and clarification questions used.)

Before questions were offered for discussion, the interviewer briefly reviewed the components of the TRIAD experience. (WyTRIAD is usually referred to simply as "TRIAD" among participants.) Also, the protocol for recording the interviews was explained and written consent forms were signed.

PRIMARY QUESTIONS:

1.   What motivated you to participate in the TRIAD? Why did you decide to participate?

2.   What did you expect the experience would be like?

3.   Were your expectations met? or how was it different?

4.   What have been the benefits of your participation?

5.   Do you feel it has had an impact?
       a.   Specifically, what?
       b.   Did the classroom experiences help in other areas?
       c.   Do you look at kids differently as a result of your experiences?
       d.   What are you doing differently as a result of your experiences?

6.   What, if any, have been your disappointments with the project as a whole?

7.   Specifically, which components were the most successful?   least successful?
              (Interviewer entered the name of each specific component into the discussion as it went along to maintain sequence and assure completeness.)

7.   Specifically, which components have you been able to continue to use? unable to continue to use? why?

(Interviewer entered the name of each specific component into the discussion as it went along to maintain sequence and assure completeness.)

8.   What have been barriers to implementation?
      a.   Overall?
      b.   In regard to specific components?

10.   What needs to change to create an ideal teaching/learning situation?

SECONDARY QUESTIONS, LOOKING BEYOND:

11.   If we were to continue and expand the TRIAD relationship, what advice do you have?
      a.   Regarding high schools?
      b.   Regarding community colleges?
      c.   To implement in high schools, what if anything would we need to change?

12.   Are any of the components of the project unrealistic?

13.   What were your greatest successes?

14.   What would you do differently if you did it again?

15.   If we continue the TRIAD next year, how would you like to be involved?

16.   Any additional advice?

B. QUESTIONS FOR THE ADMINISTRATORS

PRIMARY QUESTIONS:

1.   What is your perception of the effectiveness of the TRIAD experience?

2.   How do you feel about specific components?

3.   To what extent were your teachers able to implement the components? With which ones were they successful, and which not?

4.   What do you think are barriers to the implementation?

5.   What do you think the teachers see as barriers?

6.   How should we continue, in a way that would overcome the perceived barriers?

7.   From your standpoint, what do you think the impact of the project has been
        a.   on individual teachers?
        b.   on the schools?
        c.   on the district as a whole?

8.   Do you think it will work as well with math as it has with science?

9.   What do you see as limitations of the TRIAD?

SECONDARY QUESTIONS, LOOKING BEYOND:

10.   What would it look like if we expanded to full K-12 and also brought in the community colleges?

11.   What do you think the response of high school colleagues will be?
        a.   How can they be brought in?
        b.   How should we alter the experiences?
        c.   What components might work best? not work? problems?
        d.   How does the project/approach fit into their vision for their school?
        e.   What risks are they willing to take?
        f.   What restructuring are they willing to create?
        g.   Are they willing to look at the assessment?   curriculum?
instruction?

12.   What kind of risk are you willing to take as a district?

12.   How can we involve the present TRIAD teachers?

13.   Who should approach the community colleges and who should be the initial community college representatives?

Appendix IX.

Starting a WyTRIAD

"How to Start" list

"How to Start" list

- Know the model, the components, and the research

- Contact the district curriculum coordinator, principals, and district superintendent and make the case for the partnership

- Present about the model at the state or regional conference

- Request a meeting with the superintendent, principals, and curriculum coordinator to present and respond to questions

- Make the commitments and expectations clear

- Seek funding--leverage district funds (state and federal flow-through) with grants

PART FIVE:
References

References related to the topics incorporated into the WyTRIAD and CCM models

References
related to the topics
incorporated into the
WyTRIAD and CCM models

AAAS (1989). *Project 2061: science for all Americans.* Washington, DC: American Association for the Advancement of Science.

AAAS (1993). *Benchmarks for science literacy.* Washington, DC: American Association for the Advancement of Science.

Ackland, R. (1991). A review of the peer coaching literature. *Journal of Staff Development, 12*(1), 22-27.

Albert, E. (1978). Development of the concept of heat in children. *Science Education, 62*(3), 389-399.

Alvarez, D. (1992). Professional growth seminars encourage shared decision making, collaboration. *NAASP Bulletin, 76*(January), 70-75.

Anastos, J., & Ancowitz, R. (1987). A teacher-directed peer coaching project. *Educational Leadership*(November), 40-42.

Anderson, O. R. (1992). Some interrelationships between constructivist models of learning and current neurobiological theory, with implications for science education. *Journal of Research in Science Teaching, 29*(10), 1037-1058.

Atkins, M., & Karplus, R. (1962). Discovery or invention? *The Science Teacher, 29*(September), 45-51.

Ausubel, D. P., Novak, J. D., & Hanesian, H. (1978). *Educational psychology: a cognitive view (2 ed.).* New York: Holt, Rinehart and Winston.

Ausubel, D. P., & Robinson, F. (1969). *School learning: an introduction to educational psychology.* New York: Holt, Rinehart and Winson.

Bar, V. (1989). Introducing mechanics at the elementary school. *Physics Education, 24*, 348-352.

Batesky, J. (1991). Peer coaching. *Strategies*(June).

Bell, B., Osborne, R., & Trasker, R. (1985). Appendix A: Finding out what children think. In R. Osborne & P. Freyberg (Eds.), *Learning in Science: the implications of children's science* . Portsmouth, NH: Heinemann.

Bell, B. F. (1985). Students' ideas about plant nutrition: What are they? *Journal of Biological Education, 19*(3), 213-218.

Berg, T., & Brouwer, W. (1991). Teacher awareness of student alternate conceptions about rotational motion and gravity. *Journal of Research in Science Teaching, 28*(1), 3-18.

Bethal, L. (1988). Science teacher preparation and professional development. In A. Towsley & B. Voss, What the science education literature says about elementary school science. In L. Motz & G. Madrazo (Eds.), *Third Sourcebook for Science Supervisors* (pp. 171-180). Washington, DC: National Science Teachers Association.

Bissex, G., & Bullock, R. Eds.). (1987). *Seeing for ourselves: case study research by teachers of writing.* Portsmouth, NH: Heinemann.

Blosser, P. (1989). Disseminating research about science education. In L. Motz & G. Madrazo (Eds.), *NSSA Third Sourcebook for Science Supervisors* . Washington, DC: National Science Teachers Association.

Bonnwell, C. C., and Eison, J. A. (1991). *Active learning: creating excitement in the classroom.* Washington, D.C.: ASHE-ERIC Higher Education Report.

Bracey, G. W. (1990). Rethinking school and university roles. *Educational Leadership*(May), 65-66.

Bracey, G. W. (1989). Why so much education research is irrelevant, imitative and ignored. *American School Board Journal, 70*(7), 20-22.

Brandt, R. S. (1987). On teachers coaching teachers: a conversation with Bruce Joyce. *Educational Leadership*(February), 12-17.

Briscoe, C. (1991). The dynamic interactions among beliefs, role metaphors, and teaching practices: a case study of teacher change. *Science Education 75*, 186-199.

Brumbaugh, K. E., & Poirot, J. L. (1993, March). The teacher as researcher: presenting your case. *The Computing Teacher*, p. 19-22.

Burton, F. R. (1988). Reflections on Strickland's "Toward the Extended Professional". *Language Arts, 65*(8, December), 7765-768.

Butzow, J., & Gabel, D. (1986). Disseminating research about science education. In L. Motz & J. Madrazo G. (Eds.), *Third Sourcebook for Science Supervisors* Washington, DC: National Science Teachers Association.

Champagne, A., Gunstone, R. F., & Klopfer, L. E. (1983). Naive knowledge and science learning. *Research in Science and Technological Education, 1*(2), 173-183.

Clement, J. (1987). Overcoming students' misconceptions in physics: the role of anchoring intuitions and analogical validity. In J. Novak (Ed.), *Proceedings of the Second International Seminar on Misconceptions and Educational Strategies in Science and Mathematics* (pp. 84-97). Ithaca, NY: Cornell University.

Clement, J. (1983). Students' alternative conceptions in mechanics: a coherent system of preconceptions? In H. Helm & J. D. Novak (Eds.), *Proceedings of the International Seminar: Misconceptions in Science and Mathematics.* Ithaca, NY: Cornell University Press.

Clough, E. E., & Driver, R. (1985). Secondary students' conceptions of the conduction of heat: bringing together scientific and personal views. *Physics Education, 20*, 176-182.

Cochran-Smith, M., & Lytle, S. L. (1992). Communities for teacher research: fringe or forefront? *American Journal of Education*(May), 298-324.

Comb, A. W. (1979). *Myths in education: beliefs that hinder progress and their alternatives.* Boston: Alyn & Bacon.

Corey, S. (1954). Action research in education. *Journal of Educational Research, 47,* 375-380.

Cronin-Jones, L. (1991). Science teacher beliefs and their influence on curriculum implementation: Two case studies. *Journal of Research in Science Teaching, 28*(3), 235-250.

Cross, K. P. (1991). Every teacher a researcher, every classroom a laboratory. *Tribal College*(Spring), 7-12.

Cross, K. P., & Angelo, T. A. (1988). *Classroom assessment techniques: a handbook for faculty.* Ann Arbor: NCRIPTAL, University of Michigan.

Desrochers, C. G., & Klein, S. R. (1990). Teacher-directed peer coaching as a follow-up to staff development. *Journal of Staff Development, 11*(2), 6-10.

DeVito, A., & Krockover, G. (1990). *Creative sciencing: a practical approach.* Boston: Little, Brown, and Co.

Dreyfus, A., Jungwirth, E., & Eliovitch, R. (1990). Applying the "cognitive conflict" strategy for conceptual change--some implications, difficulties, and problems. *Science Education, 74*(5), 555-569.

Dreyfus, A., & Jungworth, E. (1989). The pupil and the living cell: a taxonomy of dysfunctional concepts about an abstract idea. *Journal of Biological Education, 21*(3), 23 (1), 49-55.

Driver, R., & Bell, B. (1985). Students' thinking and the learning of science: a constructivistic view. *School Science Review, 67*(240), 443-456.

Driver, R., & Easely, J. A. (1978). Pupils and paradigms: a review of literature related to concept development in adolescent science.

Driver, R., & Scanlon, E. (1989). Conceptual change in science. *Journal of Computer Assisted Learning, 5,* 25-36.

Duckworth, E. (1987). *The having of wonderful ideas.* New York: Teachers College Press.

Duckworth, E. (1986). *Inventing density.* Grand Forks, ND: Center for Teaching and Learning, University of North Dakota.

Duit, R. (1987). Research on students' alternative frameworks in science--topics, theoretical frameworks, consequences for science teaching. In J. Novak (Ed.), *Proceedings of the Second International Seminar on Misconceptions and Educational Strategies in Science and Mathematics* (pp. 151-162). Ithaca, NY: Cornell University.

Dyche, S., McClurg, P., Stepans, J. I., & Veath, M. L. (1993). Questions and conjectures concerning models, misconceptions and spatial ability. *School Science and Mathematics, 93*(4), 191-197.

Eaton, J., Anderson, C., & Smith, E. (1984). Students' misconceptions interfere with science learning: case studies of fifth grade students. *The Elementary School Journal, 84*(4), 365-379.

Eaton, J. F., Anderson, C. W., & Smith, E. L. (1983). When students don't know they don't know. *Science and Children, 20*(7), 6-9.

Ebert, C., & Ebert, E. (1993). An instructionally oriented model for enabling conceptual development. In J. Novak (Ed.) *Proceedings of the Third International Seminar on Misconceptons and Educational Strategies in Science and Mathematics.* Ithaca, NY: Cornell University.

Ebert, C., & McKenzie, D. (1987). Using interviewing as a teacher education technique. *Journal of Science Teacher Education, 1*(2), 27-29.

Eckstein, S. G., & Shemesh, M. (1993). Development of children's ideas on motion: impetus, the straight-down belief and the law of support. *School Science and Mathematics, 93*(6), 299-304.

Education, U. S. Dept. of. (1994). Professional development, standards, and education reform. *OERI Bulletin* (Office of Educational Research and Improvement)(Winter, 1994), 1 & 4.

Epstein, H. J. (1979). Cognitive growth and development. *Colorado Journal of Educational Research, 19*(1), 4-5.

Erickson, G. L. (1980). Children's viewpoints of heat: a second look. *Science Education, 64*(3), 323-336.

Erickson, G. L. (1979). Children's conceptions of heat and temperature. *Science Education, 63*(2), 221-230.

Erlwanger, S. H. (1975). Case studies of children's conceptions of mathematics, I. *Journal of Children's Mathematical Behavior, 1*(3), 157-283.

ESS (1971). Whistles and strings. In *Elementary Science Study* . New York: McGraw-Hill.

ESS (1971). Musical instrument recipe book. In *Elementary Science Study.* New York: McGraw-Hill.

ESS (1968). Batteries and bulbs. In *Elementary Science Study.* New York: McGraw-Hill.

ESS (1969). Pendulum. In *Elementary science study.* New York: McGraw-Hill.

ESS (1971). Heating and cooling. In *Elementary science study.* New York: McGraw-Hill.

Feher, E., & Rice, R. (1986). Shadow shapes. *Science and Children, 24*(2), 6-9.

Feher, E., & Rice, R. (1985). Development of scientific concepts through the use of interactive exhibits in a museum. *American Museum of Natural History, 28*(35-46).

Finley, F. N. (1986). Evaluating instruction: the complementary use of clinical interviews. *Journal of Research in Science Teaching, 23*(7), 635-650.

Flake, C. Kuhs, Donnelly, A., & Ebert, C. (1995). Reinventing the role of teacher: teacher as researcher. *Phi Delta Kappan* (January).

Fredette, N., & Lochhead, J. (1980). Student conceptions of simple circuits. *The Physics Teacher, 18,* 194-198.

Freeman, I. (1965). *Physics made simple.* Garden City, NY: Doubleday & Company.

Friedl, A. (1986). *Teaching science to children--an integrated approach.* New York: Random House.

Gagne, R. M. (1977). *The conditions of learning.* New York: Holt, Rinehart &Winston.

Geddis, A. (1990). What to do about "misconceptions"--a paradigm shift. In American Educational Research Association (AERA) annual meeting, . Chicago, IL: ERIC document 351186.

Gil-Perez, D., & Carrascosa, J. (1990). What to do about science "misconceptions". *Science Education*, 74(5), 531-540.

Gilbert, J., Osborne, R., & Fensham, P. (1982). Children's science and its consequences for teaching. *Science Education*, 66, 623-633.

Glesne, C. E. (1991). Yet another role? The teacher as researcher. *Action in Teacher Education, XIII*(1), 7-13.

Goodlad, J. (1975). *The dynamics of educational change.* New York: McGraw-Hill.

Goswami, D., & Stillman, P. (Eds.). (1987). *Reclaiming the classroom: teacher research as an agency for change.* Montclair, NJ: Boynton/Cook.

Guesne, E. (1984). Light. In *New trends in physics teaching* . Paris: UNESCO.

Guesne, E. (1984). Children's ideas about light. In *New Trends in Physics Teaching* Paris: UNESCO.

Gunstone, R., & Watts, M. (1985). Force and motion. In R. Driver, E. Guesne, & A. Tiberghien (Eds.), *Children's ideas in science* . Philadelphia: Open University Press.

Halloun, I., & Hestenes, D. (1985). Common sense concepts about motion. *American Journal of Physics*, 53(11), 1056-1065.

Hapkiewicz, A., & Hapkiewicz, W. G. (1993). Misconceptions in science. In National Science Teachers Association regional convention, . Denver:

Hart, K. (1984). Which comes first--length, area, or volume? *The Arithmetic Teacher, 16-18*(May), 26-27.

Hashweh, M. Z. (1986). Towards an explanation of conceptual change. *European Journal of Science Education, 8*(3), 229-249.

Hawisher, G. E., & Pemberton, M. A. (1991). The case for teacher as researcher in computers and composition studies. *The Writing Instructor*(Winter), 77-88.

Head, J. (1986). Research into "alternative frameworks": promise and problems. *Research in Science & Technological Education, 4*(2), 203-211.

Heller, P. M., & Finley, F. N. (1992). Variable uses of alternative conceptions: a case study in current electricity. *Journal of Research in Science Teaching, 29*(3), 259-275.

Herrick, M. J. (1992). Research by the teacher and for the teacher: an action research model linking schools and universities. *Action in Teacher Education, XIV*(3), 47-54.

Herrmann, B. A., Cook, K., Elliott, W., Lewis, L., & Thomas, J. (1994). Building professional contexts for learning for preservice and inservice teachers and teacher educators: reflections, issues and questions. In Columbia, SC: Unpublished paper.

Hewson, M. G., & Hewson, P. W. (1984). Effect of instruction using students' prior knowledge and conceptual strategies on science learning. *European Journal of Science Education, 6*(1), 1-6.

Hewson, P. W. (1981). A conceptual change approach to learning science. *European Journal of Science Education, 8*(3), 383-396.

Hildreth, D. (1983). The use of strategies in estimating measurements. *The Arithmetic Teacher*(January), 50-54.

Houser, N. O. (1990). Teacher-researcher: the synthesis of roles for teacher empowerment. *Action in Teacher Education, XII*(2), 55-60.

InfoEd (1995). *SPIN: Sponsored Programs Information Network.* Albany, NY 12205: InfoEd, Inc. (453 New Karner Road).

Iona, M., & Beaty, W. (1987). Reflections on refraction. *Science and Children, 25*(18-20).

Jones, C. (1989). 72 x 49. *Mathematics Teaching, 94*(March), 8-11.

Joyce, B., & Showers, B. (1982). The coaching of teaching. *Educational Leadership, 40*(1), 4-10.

Judson, H. F. (1980). *The search for solutions.* New York: Holt, Rinehart, and Winston.

Kelsay, K. L. (1991). When experience is the best teacher: the teacher as researcher. *Action in Teacher Education, XIII*(1), 14-20.

Kuehn, C., & McKenzie, D. (1987). Finding out what children know or "straight from the horse's mouth." In J. Novak (Ed.) *Proceedings of the Second International Seminar on Misconceptions and Educational Strategies in Science and Mathematics, Vol. 2,* 260-261.

Kuehn, C., & McKenzie, D. (1988). The art of the interview. *Science Scope, 11*(5), 22-23.

Kuehn, C., & McKenzie, D. (1987). Finding out what children know or "straight from thehorse's mouth." *The Georgia Science Teacher.*

Kutz, E. (1992). Teacher research: myths and realities. *Language Arts, 69*(March), 193-197.

Kyle, W., & Shymansky, J. (1988). What research says...about teachers as researchers. *Science and Children, 26*(Nov-Dec), 29-31.

Labinowicz, E. (1985). *Learning from children: new beginnings for teaching numerical thinking.* Menlo Park, CA: Addison-Wesley.

Laurel, E. G., Chapman, J., & Hoffmeyer, C. (1990-91). Employing peer coaching to support teachers. *Teacher Education and Practice*(Fall/Winter), 79-82.

Lawson, A. E. (1985). A review of research on formal reasoning and science teaching. *Journal of Research in Science Education, 22*(7), 569-617.

Lemon, D. (1990). Effective schools and the nature of their leadership. In The umbrella Reprinted from NAESP, Chapter 2 of *Principals for 21st Century Schools.*

Leyden, M. (1985). The strange silos. *Science and Children*(October), 32-33.

Lieberman, A. (1986). Collaborative research: working with, not working on. *Educational Leadership, 43*(5), 28-33.

Liedke, W. (1988). Diagnosis in mathematics: the advantages of an interview. *Arithmetic Teacher*(November), 26-29.

Liem, T. (1987). *Invitation to science inquiry.* Lexington, MA: Ginn Press.

Lightman, A., & Sadler, P. (undated). How could the earth be round? *Preprint Series No. 2580.* Harvard-Smithsonian Center for Astrophysics.

Lytle, S. L., & Cochran-Smith, M. (1992). Teacher research as a way of knowing. *Harvard Educational Review, 62*(4 Winter), 447-474.

Maeroff, G. (1988). *The empowerment of teachers: overcoming the crisis of confidence.* New York: Teachers College Press.

Maloney, D. (1990). Forces as interactions. *The Physics Teacher*(July), 386-390.

Marson, R. (1978). Motion. In TOPS (Task-Oriented Physical Science). Vista, CA: TOPS Learning Systems.

McClelland, J. A. G. (1985). Misconceptions in mechanics and how to avoid them. *Physics Education, 20,* 159-162.

McCloskey, M., Caramazza, A., & Green, B. (1980). Curvilinear motion in the absence of external forces: naive beliefs about the motion of objects. *Science, 210*(December 5), 1139-1141.

McDermott, L. (1984). Research on conceptual understanding in mechanics. *Physics Today*(July), 24-32.

McKernan, J. (1988). Teacher as researcher: paradigm and praxis. *Contemporary Education, 59*(3, Spring), 154-158.

Meier, D. (1990). Taking children's opinions seriously: a talk with Bruno Bettelheim. *Teacher Magazine*, p. 44-46.

Minstrell, J. (1982). Explaining the "at rest" condition of an object. *The Physics Teacher*(January), 10-14.

Morgan, R. L., Gustafson, K. J., Hudon, P. J., & Salzberg, C. L. (1992). Peer coaching in a preservice special education program. *Teacher Education and Special Education, 15*(4), 249-258.

NCTM. (1989). *Curriculum and evaluation standards for school mathematics.* Reston, VA: National Council of Teachers of Mathematics.

NCTM. (1994). Aichele, D. B., (Ed.). *Professional development for teachers of mathematics: 1994 yearbook.* Reston, VA: National Council of Teachers of Mathematics.

National Research Council. (1989). *Everybody counts: a report to the nation on the future of mathematics education.* Washington, DC: National Academy Press.

National Research Council (1995). *National science education standards.* (draft) National Academy of Sciences, Washington, DC: National Academy Press.

NSTA (1983). Engineering a classroom discussion. Ed. column. *Science and Children*(February), 21-22.

Neubert, G. A., & Bratton, E. C. (1987). Team coaching: staff development side by side. *Educational Leadership*(February), 29-32.

Novak, D. J. (1987). Human constructivism: towards a unity of psychological and epistemological meaning making. In J. Novak (Ed.), *Second International Seminar on Misconceptions and Educational Strategies in Science and Mathematics Education.* Ithaca, NY: Cornell University.

NSDC (1995). *NSDC'S standards for staff development: elementary level edition.* Oxford, OH: National Staff Development Council.

NSDC (1994). *NSDC'S standards for staff development: middle level edition.* Oxford, OH: National Staff Development Council.

NSDC (1995). *NSDC'S standards for staff development: high school level edition.* Oxford, OH: National Staff Development Council.

Nussbaum, J., & Novak, J. D. (1976). An assessment of children's conceptions of the earth using structured interviews. *Science Education, 60,* 535-550.

Nussbaum, J., & Novick, S. (1981). Creating cognitive dissonance between students' preconceptions to encourage individual cognitive accommodation and a group cooperative construction of a scientific model. In AERA Annual Convention., Los Angeles, CA:

Osborne, R. (1984). Children's dynamics. The Physics Teacher(November), 504-508.

Osborne, R., & Freyberg, P. (1985). *Learning in science: the implications of children's science.* London: Heinemann.

Osborne, R., & Wittrock, M. (1983). *Learning science: a generative process.* Science Education, 67, 490-508.

Oxenhorn, J. (1982). *Pathways to science: sound and light.* New York: Globe Book Company.

Oxenhorn, J., & Idelson, M. (1982). *Pathways in science, the forces of nature.* New York: Globe Book Co.

Peck, D. M., Jencks, S. M., & Connell, M. L. (1989). Improving instruction through brief interviews. *Arithmetic Teacher*(November), 15-17.

Phillips, M. D., & Glickman, C. D. (1991). Peer coaching: developmental approach to enhancing teacher thinking. *Journal of Staff Development*, 12(2), 20-25.

Piaget, J. (1969). *The child's conception of the world.* Totowa, NJ: Littlefield, Adams & Co. (Originally published in 1929.).

Piaget, J. (1964). Cognitive development in children: development and learning. *Journal of Research in Science Teaching, 2,* 176-186.

Poirot, J. L. (1992, August/September). Assessment and evaluation of technology in education--the teacher as researcher. *The Computing Teacher,* p. 9-10.

Porter, A., & Brophy, J. (1988). Synthesis of research on good teaching: insights from the work of the Institute for Research on Teaching. *Educational Leadership*(May).

Posner, G., Strike, K., Hewson, P., & Gertzog, W. (1982). Accommodation of a scientific conception: toward a theory of conceptual change. *Science Education, 66*, 211-227.

Posner, G. J., & Gertzog, W. A. (1982). The clinical interview and the measurement of concept changes. *Science Education, 66*(2), 95-200.

Powney, J., & Watts, D. M. (1987). *Interviewing in educational research.* London: Routledge & Kegan Paul.

Project AIMS (1991). *Mostly magnets.* Fresno, CA: AIMS Educational Foundation.

Raney, P., & Robbins, P. (1989). Professional growth and support through peer coaching. *Educational Leadership*(May), 35-38.

Renner, J. W., & Marek, E. A. (1988). *The learning cycle and elementary school science teaching.* Portsmouth, NH: Heinemann Educational Books, Inc.

Roberts, J. (1991). Improving principals' instructional leadership through peer coaching. *Journal of Staff Development, 12*(4), 30-33.

Rogan, J. M. (1988). Development of a conceptual framework of heat. *Science Education, 72*(1), 103-113.

Rowe, M. (1988). Science education: a framework for decision-makers. In L. Motz & J. Madrazo G. (Eds.), *Third Sourcebook for Science Supervisors.* Washington, DC: National Science Teachers Association.

Rowell, J. A., & Dawson, C. (1977). Teaching about floating and sinking: further studies toward closing the gap between cognitive psychology and classroom practice. *Science Education, 61*(4), 527-540.

Rudduck, J., & Hopkins, D. (1985). Research as a basis for teaching: readings from the work of Lawrence Stenhouse. London: Heinemann Educational Books, Inc.

Rutherford, J. (1995). Hyphen-based reform. *2061 Today* (American Association for the Advancement of Science), 5(1), 7.

Sadanand, N., & Kess, J. (1990). Concepts in force and motion. *The Physics Teacher*(November), 530-533.

Saigo, B. (1994). Getting $tarted on grant$: money to do what you want to do. Presentation at NSTA Area Convention, Portland, Oregon, October 13-15.

Schaefer, R. J. (1967). *School as a center of inquiry.* New York: Harper and Row.

Scott, D. (1986). *The physics of vibrations and waves.* Columbus, OH: Merrill Publishing Co.

Shalaway, L. (1990). Tap into teacher research. *Instructor*(August), 34-38.

Shayer, M., & Wylam, H. (1981). The development of the concepts of heat and temperature in 10-13 year-olds. *Journal of Research in Science Teaching, 18*(5), 419-434.

Shepherd, D., & Renner, J. W. (1982). Student understandings and misunderstandings of states of matter and density changes. *School Science and Mathematics, 82*(8), 650-665.

Shipstone, D. (1988). Pupils' understanding of simple electrical circuits. *Physics Education, 23*, 92-96.

Shipstone, D. (1985). Electricity in simple circuits. In R. Driver, E. Guesne, & A. Tiberghien (Eds.), *Children's ideas in science* (pp. 33-51). Philadelphia: Open University Press.

Showers, B. (1985). Teachers coaching teachers. *Educational Leadership,* 42(7, April), 43-49.

Schrodinger, E. (1958). *Mind and matter.* Boston: Cambridge University Press.

Shuell, T. J. (1987). Cognitive psychology and conceptual change: implications for teaching science. *Science Education, 71*(2), 239-250.

Snir, J. (1989). Making waves: a simulation and modeling computer-tool for studying wave phenomena. *Journal of Computers in Mathematics and Science Teaching, 8*(4), 48-53.

Sparks, G. M., & Bruder, S. (1987). Before and after peer coaching. *Educational Leadership*(September), 54-57.

Stenhouse, L. (1988). Artistry and teaching: the teacher as focus of research and development. *Journal of Curriculum and Supervision, 4*(1), 43-51.

Stenhouse, L. (1975). *An introduction to curriculum research and development.* London: Heinemann Educational Ltd.

Stepans, J. (1992). Using the pendulum to illustrate the conceptual change teaching strategy. Unpublished document. Laramie, WY: University of Wyoming.

Stepans, J. (1994). *Targeting students' science misconceptions: physical science activities using the conceptual change model.* Riverview, FL 33569: Idea Factory, Inc. (10710 Dixon Drive).

Stepans, J. (1990). Balancing. Unpublished document. Laramie, WY: University of Wyoming.

Stepans, J. (1991). Developmental patterns in students' understanding of physics concepts. In S. Glynn, R. Yeany, & B. Britton (Eds.), *The psychology of learning science* . Hillsdale, NJ: Lawrence Erlbaum, Publishers.

Stepans, J. (1991). Will it mix, sink or float? *School Science and Mathematics, 91*(5), 218-220.

Stepans, J. (1988). What are we learning from children about teaching and learning? *Journal of Natural Inquiry, 2*(2), 9-18.

Stepans, J. (in review). Learning about density. *Science and Children.*

Stepans, J. (1990). On the feasibility of research collaboration between higher education faculty and public school teachers. *Researcher*), 6(2), 3-5. Northern Rocky Mountain Educational Research Association.

Stepans, J. (1989). A partnership for making research work in the classroom. *Researcher* , 6(1), 3-5. Northern Rocky Mountain Educational Research Association

Stepans, J., Beiswenger, R. E., & Dyche, S. (1986). Misconceptions die hard. *Science Teacher, 53*(6), 65-69.

Stepans, J., Dyche, S., & Beiswenger, R. (1988). The effect of two instructional models in bringing about a conceptual change in the understanding of science concepts by prospective elementary teachers. *Science Education, 72*(2), 185-195.

Stepans, J., Miller, K., & Willis, C. (1992). Teacher, administrator, and university educator form a triad. *Journal of Rural and Small Schools, 52* (Spring)), 38-41.

Stepans, J. I., and Ebert, C. (1993). A partnership for changing the school from within. In AETS Annual Convention, . Charleston, SC:

Stepans, J. I. (1987). Designing math and science lessons. Unpublished document. Laramie, WY: University of Wyoming.

Stepans, J. I. (1985). Biology in elementary schools: children's conceptions of life. *American Biology Teacher 47*, 222-225.

Stepans, J. I., & Saigo, B. W. (1993). Barriers which may keep teachers from implementing what we know about identifying and dealing with students' science and mathematics misconceptions. In J. Novak (Ed.), *Third International Seminar on Misconceptions and Educational Strategies in Science and Mathematics* . Ithaca, NY: Cornell University.

Stepans, J. I., Saigo, B. W., & Ebert, C. (1995). *Changing the classroom from within: an in-service model based on the Wyoming TRIAD (WyTRIAD)*. Montgomery, AL 36124: Saiwood Biology Resources (P. O. Box 242141).

Stepans, J. I., & Veath, M. L. (1990). On research. *Science Scope, 33*(Nov./Dec.), 52.

Stepans, J. I., & Veath, M. L. (1994). How do students really explain changes in matter? *Science Scope, 17*(8), 31-35.

Stepans, J. I., & Veath, M. L. (1990). The use of models in a science classroom. *Science Scope, 14*(3), 33.

Stepans, J. I., & Kuehn, C. (1985). Children's conception of weather. *Science and Children 23*, 44-47.

Strauss, M. J., & Levine, S. H. (?1985-86). Symbolism, science and developing minds. *Journal of College Science Teaching, 15*(3), 190-195.

Strickland, D. S. (1988). Reflections on Burton's reflections. *Language Arts, 65*(8, December), 769-770.

Strickland, D. S. (1988). The teacher as researcher: toward the extended professional. *Language Arts, 65*(8, December), 754-764.

Strike, K., & Posner, G. (1985). A conceptual change view of learning and understanding. In L. West & A. Pines (Eds.), *Cognitive structure and conceptual change* . Orlando, FL: Academic Press.

Taagepera, M., & Lewis-Knapp, Z. (Eds.). (1989). *Science demonstration lessons.* Irvine, CA: University of California, Irvine.

Terry, C., Jones, G., & Hurford, W. (1985). Children's conceptual understanding of forces and equilibrium. *Physics Education, 20*, 162-165.

Tobias, S. (1990). *They're not dumb, they're different: stalking the second tier.* Tucson, AZ: Research Corporation.

Tobin, K. (1990). Research on science laboratory activities: in pursuit of better questions and answers to improve learning. *School Science and Mathematics, 90*(5), 403-412.

Tobin, K., Capie, W., & Bettencourt, A. (1988). Active teaching for higher cognitive learning in science. *International Journal of Science Education, 10*(1), 17-27.

Tomlinson, P. (1989). Having it both ways: hierarchical focusing as research interview method. *British Educational Research Journal, 15*(2), 155-176.

Trumbull, D. J., & Lack, M. J. (1991). Learning to ask, listen, and analyse: using structured interviewing assignments to develop reflection in preservice science teachers. *International Journal of Science Education, 13*(2), 129-142.

Walton, E. (1988). How to support your science program: an introduction. In L. Motz & J. Madrazo G. (Eds.), *Third Sourcebook for Science Supervisors* . Washington, DC: National Science Teachers Association.

Wandersee, J. H. (1986). Can the history of science help science educators anticipate students' misconceptions? *Journal of Research in Science Teaching, 23* (7):581-597.

Watson, B., & Konicek, R. (1990). Teaching for conceptual change: confronting children's experience. *Phi Delta Kappan*(May).

Watts, D. (1985). Student conceptions of light: a case study. *Physics Education, 20*, 183-187.

Watts, D. M., & Zylbersztajn, A. (1981). A survey of some children's ideas about force. *Physics Education, 16*, 360-365.

Watts, M., & Ebbutt, D. (1987). More than the Sum of the Parts: research methods in group interviewing. *British Educational Research Journal, 13*(1), 25-34.

Watts, M. D. (1983). A study of schoolchildren's alternative frameworks of the conception of force. *European Journal of Science Education, 5*, 217-230.

West, L. H. T., & Pines, L. (1984). An interpretation of research in "conceptual understanding" within a source-of-knowledge framework. *Research in Science Education, 14*, 47-56.

Whitaker, R. J. (1983). Aristotle is not dead: student understanding of trajectory motion. *American Journal of Physics, 51*, 352-357.

Williamson, L. S., & Russell, D. S. (1990). Peer coaching as a follow-up to training. *Journal of Staff Development, 11*(2), 2-4.

Wolcott, L. (1991, Teacher, researcher. *Teacher Magazine*, p. 28-29.

Wolfinger, D. (1984). *Teaching science in the elementary school: content, process, and attitude.* Boston: Little, Brown, & Co.

Wright, E., & Perna, J. (1992). Reaching for excellence: a template for biology instruction. *Science and Children 30*(2), 35.

Zaslavsky, C. (1989). People who live in round houses. *The Arithmetic Teacher*(September), 18-21.

Zietman, A., & Clement, J. (1990). Using anchoring conceptions and analogies to teach about levers. In AERA annual meeting, . Boston, MA.